MATHEMATICAL RESEARCH | MATHEMATISCHE FORSCHUNG

# Parametric Integer Optimization

B. Bank / R. Mandel

Band 39

Akademie-Verlag Berlin

In this series original contributions of mathematical research in all fields are contained, such as

— research monographs
— collections of papers to a single topic
— reports on congresses of exceptional interest for mathematical research.

This series is aimed at promoting quick information and communication between mathematicians of the various special branches.

In diese Reihe werden Originalbeiträge zu allen Gebieten der mathematischen Forschung aufgenommen wie

— Forschungsmonographien
— Sammlungen von Arbeiten zu einem speziellen Thema
— Berichte von Tagungen, die für die mathematische Forschung besonders aktuell sind.

Die Reihe soll die schnelle Information und gute Kommunikation zwischen den Mathematikern der verschiedenen Fachgebiete fördern.

---

Manuscripts
in English and German comprising at least 100 pages and not more than 500 pages can be admitted to this series. With respect to a quick publication the manuscripts are reproduced photomechanically. Authors who are interested in this series please turn directly to the 'Akademie-Verlag'. Here you will get more detailed information about the form of the manuscripts and the modalities of publication.

Manuskripte
in englischer und deutscher Sprache, die mindestens 100 Seiten und nicht mehr als 500 Seiten umfassen, können in diese Reihe aufgenommen werden. Im Interesse einer schnellen Publikation werden die Manuskripte auf fotomechanischem Weg reproduziert. Autoren, die an der Veröffentlichung entsprechender Arbeiten in dieser Reihe interessiert sind, wenden sich bitte direkt an den Akademie-Verlag. Sie erhalten dort genauere Informationen über die Gestaltung der Manuskripte und die Modalitäten der Veröffentlichung.

B. Bank / R. Mandel

# Parametric Integer Optimization

Mathematical Research · Mathematische Forschung

Wissenschaftliche Beiträge
herausgegeben von der
Akademie der Wissenschaften der DDR
Karl-Weierstraß-Institut für Mathematik

Band 39

# Parametric Integer Optimization

by

B. Bank

and

R. Mandel

# Parametric Integer Optimization

by

Bernd Bank
and
Reinhard Mandel

Akademie-Verlag Berlin 1988

Autoren:

Doz. Dr. sc. nat. Bernd Bank
Dr. sc. nat. Reinhard Mandel
Humboldt-Universität zu Berlin
Sektion Mathematik

---

Die Titel dieser Schriftenreihe werden vom Originalmanuskript der Autoren reproduziert.

ISBN 3-05-500398-5

ISSN 0138-3019

Erschienen im Akademie-Verlag Berlin, DDR-1086 Berlin, Leipziger Str. 3-4

Lizenznummer: 202·100/425/87
Printed in the German Democratic Republic
Gesamtherstellung: VEB Kongreß- und Werbedruck, 9273 Oberlungwitz
Lektor: Dr. Reinhard Höppner
LSV 1085
Bestellnummer: 763 763 1 (2182/39)
02000

## Preface.

The subject matter of this volume is concerned with optimization problems in which some of the data involved are seen as parametric quantities and some or all of the variables are required to be integer.
The characterization of the parameter-dependent behaviour of the feasible region, the extreme value, and the set of optimal points elaborated here is of great importance both in theory and practice. This monograph is intended to be a first comprehensive contribution to the theory of parametric nonlinear and linear integer programming and, it is directed to mathematicians as well as to researchers, practical workers and students who are familiar with the basic knowledge of mathematical programming.
The book mainly contains research results obtained by the authors during the last five years and reflects the important contributions to the theory of parametric integer programming due to C.E. Blair and R.G. Jeroslow and that of R. Hansel to the quadratic integer case.
Our work was essentially influenced by fruitful discussions with several colleagues, including, above all, Professor F. Nožička, of the division Mathematische Optimierung at the Sektion Mathematik of the Humboldt-Universität zu Berlin, and stimulated by the collaboration with a research group at the Lomonossow University Moscow headed by E.G. Belousov. In particular, the authors would like to express their deep gratitude to D. Klatte and B. Kummer for their insightful discussions and helpful comments, and to J. Guddat, who steadily encouraged our interest in this subject and writing this monograph.
We have become greatly indebted to J. Kerger, Ch. Reimann and M. Willenberg for the assistance in preparing the final version of the manuscript, to S. Schmidt for her careful typing, and we would like to thank R. Höppner and G. Reiher of the Akademie-Verlag for their patience and support.
Finally, we welcome all comments, criticisms and suggestions, which should be directed to the authors:

Humboldt-Universität Berlin
Sektion Mathematik
PSF 1297
Berlin, DDR
1086

Berlin, August 1986

Bernd Bank
Reinhard Mandel

Contents

## 1. Introduction

Stability theory and the parametrization of optimization problems are of great importance both in theory and applications. Optimization models of real-life problems, like any mathematical model of such problems, only touch, in a certain approximative sense, the surface of the relations in the reality. For instance, if the functional dependences in the optimization problem are sufficiently well seized, then the applicability of the model essentially depends on the accuracy of the initial data and, a possible methodology to treat this problem is obtained by parametrization of the data in the optimization models. On the other hand, a parametrization of the data in an optimization problem reflects a typically mathematical framework successfully applied in almost all branches of mathematics and their applications. During the last decade a remarkable success has been made on stability analysis and parametrization of the models in nonlinear programming. For a comprehensive survey we refer the reader to the monographs: BANK/GUDDAT/KLATTE/KUMMER/TAMMER (1982) (for shortness in the following: BANK et al. (1982)), BROSOWSKI (1982), DINKELBACH (1969), DONTCHEV (1983), FIACCO (1983), GAL (1973,1979), NOŽIČKA/GUDDAT/HOLLATZ/BANK (1974) and the Mathematical Programming Studies 10 (1979), 19(1982) and 21(1984).
But, if the parametric nonlinear program contains integer requirements on the variables (or on some of them), then, up to now, only a few results can be listed. One reason for this situation is that one has to handle the difficulties of nonlinear programming as well as those which arise by the fact that just little is known about (mixed-) integer solutions of nonlinear equations and inequalities and their structure.

In this book we focus our interest on parametric (nonlinear mixed-) integer optimization problems which, in general, can be written in the form (used symbols and notations are explained at the end of the Introduction)

$$P(\lambda) \quad \inf\{f(x,\lambda) \;/\; x\in M(\lambda),\ x_1,\dots,x_s \text{ integer}\},\ \lambda\in\Lambda$$

where the parameter set $\Lambda$ is a subset of $R^m$, $f\colon R^n \times \Lambda \longrightarrow R$ is the objective function and the constraint set $M(\lambda)$, for every $\lambda\in\Lambda$, is a subset of $R^n$. (As usually $P(\lambda)$ is called a mixed-integer, a pure integer or "continuous" optimization problem if $0<s<n$, $s = n$ or $s = 0$, respectively.)
The behaviour of the problem $P(\lambda)$ is essentially characterized by the properties of the <u>constraint set-mapping</u>

$$M\colon\Lambda \longrightarrow 2^{R^n} \text{ given by } M(\lambda)\subset R^n,$$

the <u>value function</u>

$$\varphi\colon\Lambda \longrightarrow R\cup\{+\infty,-\infty\} \text{ defined by } \varphi(\lambda) = \inf_{x\in M(\lambda)} f(x,\lambda)$$

and the <u>optimal set-function</u>

$$\psi\colon\Lambda \longrightarrow 2^{R^n} \text{ defined by } \psi(\lambda) =\{x \in M(\lambda) \;/\; f(x,\lambda) = \varphi(\lambda)\}.$$

For the problem $P(\lambda)$ it is desirable that these mappings possess certain continuity or Lipschitzian properties (of global or local kind). The semi-continuity analysis of the above mappings, which will play a central role in our considerations, permits answers to the questions:
1. Does the accuracy of the solutions and extremal values increase with the degree of approximation of the problem data $\lambda$ ? 2. Can a solution of $P(\lambda^0)$ obtained for a fixed $\lambda^0$ be considered an approximative solution for the problems $P(\lambda)$ occurring under small perturbation of the data $\lambda$? We shall use the notation 'stability' of $P(\lambda)$, which is not of a uniform meaning, in the literature to optimization in order to announce that certain semicontinuity properties of the multifunctions above are ensured.

First papers related to parametric (mixed-) integer linear optimization problems were published in the sixties by GOMORY (1965), FRANK (1967) and NOLTEMEIER (1970). The further development of this direction was mainly influenced by the contributions of MEYER (1974,1975,1976), BANK (1977), GEOFFRION/NAUSS (1977) and BLAIR/JEROSLOW (1977,1979,1982). While the first three papers were essentially oriented to solution methods for one- and more-parametric linear problems with parametrized linear objective functions and right-hand sides of the linear constraints, MEYER was the first who introduced the multifunctional considerations into the investigation of parametric integer problems in order to analyse the existence of solutions and the stability behaviour of such models. A comprehensive global and local analysis of the constraint set-mapping, the value function and the optimal set-function for (mixed-) integer linear problems with variable objective functions and right-hand sides of the constraints is due to BANK. BLAIR/JEROSLOW obtained an important quantitative characterization of (mixed-) integer linear problems with variable right-hand sides in terms of a kind of Lipschitzian properties for the value function and the optimal set-function. In 1982 BLAIR/JEROSLOW could extend their results and show that the value function of pure integer linear problems with variable right-hand sides can be identified with thr so-called GOMORY-functions and, thus they obtained an algorithmic insight into the character of these value functions. All these results mentioned before can be considered to be a comprehensive theoretical background for (mixed-) integer linear problems (with a fixed constraint matrix; in the case that the constraint matrix is parametrized, no deep results are known up to now).

Numerical procedures for the analysis of parameter-dependent (mixed-) integer linear optimization problems are considered from different points of view in several papers, e.g. FRANK (1967), NOLTEMEIER (1970), MARSTEN/MORIN (1977), SHAPIRO (1977), HOLM/KLEIN (1978), SEELÄNDER (1980, 1980a).
Similar as in the case of 'continuous' optimization problems the para-

metric framework related with (mixed-) integer problems shows a great importance for the development of a duality theory and solution methods. Remarkable progress in the (mixed-) integer linear case can be realized in the subadditive duality approach, cutting-plane theory, for penalty functions and partitioning procedures, see e.g. GOMORY (1965,1969), CHVÁTAL (1973), BURDET/JOHNSON (1974), JEROSLOW (1978,1979), SCHRIJVER (1979), BACHEM/SCHRADER (1978), BLAIR/JEROSLOW (1981,1982), BENDERS (1962), BANK/MANDEL/TAMMER (1979), TIND/WOLSEY (1981).
For postoptimality and sensitivity analyses in (mixed-) integer linear optimization we refer the reader to the articles due to GEOFFRION/NAUSS (1977) and WOLSEY (1981). Various papers deal with parametric combinatorial problems (0-1 problems), which will not be treated in this book; for references see the BIBLIOGRAPHIES I., II., III.

In contrast to the situation just analysed rather little is known about nonlinear problems in parametric (mixed-) integer programming. The stability analyses explored in the articles by RADKE (1975) and ALLENDE (1980) are obtained under compactness requirements imposed on the constraint sets of the nonlinear problems. The basis of our considerations of parametric (mixed-) integer nonlinear optimization problems is given by the contributions of HANSEL (1980), BANK/HANSEL (1984), MANDEL (1985) and BANK/BELOUSOV/MANDEL/ČEREMNYCH/SHIRONIN (1986) (cited in the following as BANK/BELOUSOV et al. (1986)), where no compactness for the constraint sets is assumed.
The main part of our considerations of parametric (mixed-) integer nonlinear programming is related with the stability analysis of the general class of problems

$$P(p,b) \quad \inf\{f(x) + p^T x \;/\; x \in M(b), x_1,\dots,x_s \text{ integer}\},\quad p \in R^n,\ b \in R^m,$$

where the fixed function $f : R^n \longrightarrow R$ is quadratic or convex polynomial (without a linear part in both cases) and the constraint sets $M(b)$ are of the form

$$M(b) = C(b) + V \subset R^n,\quad b \in R^m, \tag{1}$$

and given by a compact-valued upper semicontinuous multifunction $C$ and a fixed convex polyhedral cone $V$. Further explorations are carried out for problems $P(p,b)$ with constraint sets of the implicit form

$$M(b) = \{x \in R^n \;/\; p_j(x) \leq b_j,\ j=1,\dots,m\},\quad b \in R^m, \tag{2}$$

where $p_j(x)$, $j=1,\dots,m$, are quasiconvex polynomial functions on $R^n$. These latter constraint sets belong, under a certain assumption ((CPC), see below), to the class defined by (1).
The following three aspects of a stability theory for $P(p,b)$ are treated in the book:
- Existence and stability of feasible points,
- Semicontinuity properties of the value functions of $P(p,b)$, their

level-sets and optimal sets,

- Existence of optimal points.

Parametric (mixed-) integer linear programming is a special case of the general problems just explained, for which additional, more sophisticated results (quasi-Lipschitzian properties) can be derived. Further, for this class of problems several applications of stability results to duality and solution concepts are demonstrated.

Now we elucidate the arrangement and the contents of the single chapters.
Chapter 2. After defining the different semicontinuities used for the multifunctions in our context we present some known basic results for parameter-dependent constraint sets in $R^n$ without integer requirements which are fairly general and, in particular, closely related to (mixed-) integer nonlinear programming cases. Some of the theorems are valid in more general spaces (see BANK et al. (1982)), our limitation to $R^n$ is natural and allows certain simplifications. The considerations of this chapter will be completed by the investigation of constraint sets M(b) described by quasiconvex polynomial inequalities (Chapter 4.) and linear constraints (Chapter 4. and 8.) where also Lipschitzian properties of the corresponding multifunctions are treated.

Chapter 3. Here, the aim is to give a survey on the known qualitative stability results for quite general nonlinear optimization problems in $R^n$ in a form associated with (mixed-) integer problems (see BANK et al. (1982)).
This presentation is carried on by the results in Chapter 4. for (quasi) convex polynomial optimization problems as well as for the case $s = 0$ (i.e. no integer variables) in the (mixed-) integer problems P(p,b) from Chapter 6. and P(b) from Chapter 8.

Chapter 4. Firstly, we derive basic properties of quasiconvex polynomials and constraint sets M(b) of the form (2) (where we partially follow a similar way proposed by BELOUSOV (1977) for convex polynomial functions). Taking into account results and techniques from Chapter 2. these properties enable us to examine the constraint sets M(b) in view of the continuity of the corresponding multifunction M, the representability of M(b) as a sum of a compact set C(b) and the recession cone V and the identification of such inequalities $p_j(x) \leq b_j$ which only are essential for the existence of (mixed-) integer points in M(b).

An important result with respect to the stability study for (mixed-) integer problems P(p,b) in Chapter 6. consists in the fact that the constraint sets M(b) from (2) belong to the class defined by (1) if at least one set $M(b^0)$ is representable in the form (1) (condition (CPC)). Moreover, in (1) a compact-valued multifunction C which is continuous on the effective domain of M if the condition (CPC) is satisfied can be chosen. The proof of this result, which makes use of an idea of KUMMER

(1977) for the included special case of linear constraints $p_j(x) \leq b_j$, permits a linear programming proof of a version of HOFFMAN's (1952) well-known Lipschitzian result for the mapping M. Further, in this linear case we obtain that there is even a Lipschitzian multifunction C allowing the representation (1), which is a result of importance for our considerations in Chapter 8.
In order to be able to characterize (mixed-) integer points (Chapter 5.) in sets M(b) of the form (2) we investigate the so-called stable mapping $M_{st}$ associated to M which respects only the inequalities $p_j(x) \leq b_j$ essential for the existence of such points.
Summarizing stability results for quasi-convex polynomial optimization problems we complete this chapter.

Chapter 5. is concerned with (mixed-) integer constraint sets

$$G(b) = \{x \in R^n \ / \ x \in M(b),\ x_1, \ldots, x_s \text{ integer}\},\ b \in R^m, \qquad (3)$$

First we discuss the 'uniform distribution' of (mixed-) integer points in subsets of the space $R^n$ which have a representation as the sum of a bounded set and a convex cone. We extend some results of BELOUSOV (1977) from the pure integer case to the mixed-integer one, where we restrict ourselves to such questions necessary for our further considerations; a comprehensive study of the distribution of integer points in convex sets in $R^n$ can be found in SHIRONIN (1980). For the case that M(b) in (3) has the form (1) we can characterize the structure of the sets G(b) and their convex hulls and show that the corresponding multifunctions are upper semicontinuous on their common domain if the included polyhedral cone V from (1) has a system of mixed-integer generators (condition (MIG)).
The second part of this chapter deals with sets G(b) where M(b) is given by quasi-convex polynomial inequalities (i.e. M(b) from (2)). In particular, we present conditions for the existence of (mixed-) integer points in sets M(b). Under the assumptions (CPC) and (MIG) we can expose subsets of the effective domain of the multifunction G on which this mapping shows continuity. A result due to TARASOV/KHACHIYAN (1980) giving an estimate of the norm of at least one point belonging to G(0) (where M(b) from (2) is described by convex polynomial functions with integer coefficients) completes this part of Chapter 5., which can be considered a first more comprehensive study of (mixed-) integer quasiconvex polynomial inequality systems.

The third section is related with the particular linear diophantine case intensively studied during the seventies, where we give a quasi-Lipschitzian analysis of the multifunction G and related compact-valued mappings. The chapter terminates with a characterization of the vertices of the convex hull of G(b) (where M(b) has the standard form of linear programming with integer data), which is similar to that for a feasible basic point in linear programming.

Chapter 6. contains the stability analysis of the parametric (mixed-) integer problems P(p,b) (as introduced above). We characterize the continuity behaviour of the value functions $\varphi$, the optimal set-functions $\psi$ and the $\varepsilon$-optimal set-functions $\tilde{\psi}$. Most continuity results for both kinds of the fixed part f(x) in the objective function are derived for the general class of constraint sets given by (1). The more special case of constraint sets M(b) defined in (2) where the condition (CPC) is fulfilled has only to be considered in order to guarantee the upper semicontinuity of the $\varepsilon$-optimal set-function $\tilde{\psi}$ on certain subsets of its domain. Already the following simple example of a parametric linear integer problem shows that neither the value function is upper semicontinuous nor the optimal set-function is upper semi-continuous if the constraint set-mapping is not continuous.

$$P(b) \quad \inf\{-x \;/\; 0 \leq x \leq b,\; x \text{ integer}\},\; b \in R.$$

For $\{b_t\}$ with $b_t \underset{Df}{=} 1 - \frac{1}{t} \longrightarrow b_o = 1$ one has $\varphi(b_t) = 0$ and $\varphi(b_o) = -1$ and, further, $\psi(b_t) = \{0\}$ and $\psi(b_o) = \{1\}$.
The level set-mapping N corresponding to the problem P(p,b) defined by

$$N(p,b,a) = \{x \in R^n \;/\; x \in G(b), f(x)+p^T x \leq a\}$$

plays an important role for the derivation of results as well as by its own rights in a characterization of the stability behaviour of optimization problems, which is in particular true for (mixed-) integer problems. HANSEL (1980) was the first who used the level sets in the analysis of parametric quadratic (mixed-) integer problems and our proof is a modified version of HANSEL's one corresponding to our case.
For the problems P(p,b) in Chapter 6. the question for the existence of optimal points is also answered. This subject of general interest is more intensively investigated in Chapter 7. Obviously, a (mixed-) integer problem

$$\inf\{f(x) \;/\; x \in M,\; x_1,\ldots,x_s \text{ integer}\}, \tag{4}$$

where $M \subset R^n$ and $f : M \longrightarrow R$, has an optimal point if the trivial necessary conditions:

$$\text{(NC)} \qquad \begin{array}{l} M_s = \{x \in M \;/\; x_1,\ldots,x_s \text{ integer}\} \neq \emptyset, \\ \inf\{f(x) \;/\; x \in M_s\} > -\infty \end{array}$$

are satisfied. We point out for which classes of problems (4) the conditions (NC) are also sufficient for the existence of an optimal point (if only fairly weak additional assumptions are allowed). The results to this question for (mixed-) integer problems included here reflect that a level of knowledge which is comparable to that known in nonlinear optimization has been reached.

Chapter 8. Here we treat (mixed-) integer linear optimization problems

with variable right-hand sides and a rational constraint matrix from a quantitative stability point of view, i.e. we are looking for estimates for the value function $\varphi(b)$ and the optimal set-function $\Psi(b)$, and we demonstrate their quasi-Lipschitzian behaviour.
The investigations of Chapter 5. enable us to present a unified approach, which permits an extension of results due to BLAIR/JEROSLOW (1977,1979)

In Chapter 9. the importance of such a quantitative parametric concept in the theory of (mixed-) integer programming is shown, establishing on its basis appropriate penalty functions.

Another application of the parametric framework we discuss in relation to the development of partitioning methods based on a general BENDERS decomposition principle. The relations between the value function of optimization problems with variable right-hand sides and duality concepts known from "continuous" optimization can also be observed for the problems treated in Chapter 8. This duality problem is reviewed on the basis of the papers by JEROSLOW (1978) and BACHEM/SCHRADER (1978). Further, we give a short outline on the identification between the so-called Gomory functions and the value functions of parametric pure integer problems with variable right-hand sides and integer data due to BLAIR/JEROSLOW (1982).

The majority of the basic concepts employed in this monograph are derived from convex analysis and parametric optimization, here for the main part we follow the books of ROCKAFELLAR (1970) and BANK et al. (1982).

## List of Symbols and Notations

| | |
|---|---|
| $\subset$ ; $\cup$ ; $\cap$ | set inclusion; union; intersection; |
| $a \Longrightarrow b$ | a implies b; |
| $a \Longleftrightarrow b$ | a if and only if b; |
| * | the asterisk after 'without loss of generality' signifies that a proposition is true subject to the choice of an infinite subsequence; |
| $\forall$ ; $\exists$ | for all; there exists; |
| $\emptyset$ | empty set |
| $\underset{Df}{=}$ | equal by definition; |
| $\{x / \dots\}$ | set of all x such that ...; |
| R | field of the real numbers; |
| $R_+$ | set of all nonnegative real numbers; |
| $R^n$ | n-dimensional Euclidean space; |
| $[a,b]$; $(a,b)$ | closed interval; open interval |
| $[x]$ | integer part of the real number x; |
| $\ulcorner x \urcorner$ | smallest integer number greater than x; |
| $2^C$ | power set of the set C; |
| $A^T$; rank A | transpose of the matrix A; rank of A; |
| $z = (x,y)$ | decomposition of a vector z into subvectors x and y; |
| $A_s$ | $\underset{Df}{=} \{x \in A / x_1, \dots, x_s \text{ integer}\}$, (mixed-) integer points in A, $0 \leq s \leq n$; |
| inf | infimum; |
| sup | supremum; |
| lim | limit |
| $\underline{\lim}$; $\overline{\lim}$ | lower limit; upper limit; |
| $f : X \longrightarrow Y$ | function f from X into Y; |
| $F : X \longrightarrow 2^Y$ | multifunction from X into Y; |

If A,B $R^n$, $\varepsilon > 0$ and $x,y \in R^n$, then we denote by

| | |
|---|---|
| $x \leq y$ | $x_i \leq y_i$, $i=1,\dots,n$; |
| $s(x,y)$ | line segment between x and y; |
| $x^Ty$ | scalar product of x and y; |
| $\|x\|$ | Euclidean norm of x; |
| $d(x,y)$ | $\underset{Df}{=} \|x-y\|$ ,Euclidean distance between x and y |

| | |
|---|---|
| $d(x,A)$ | $\underset{Df}{=} \inf\{d(x,y) \,/\, y \in A\}$, distance between the point x and the set A; |
| $d(A,B)$ | $\underset{Df}{=} \inf\{\varepsilon \mid A \subset U_\varepsilon B,\ B \subset U_\varepsilon A\}$, distance between the sets A and B (contexts avoid misunderstandings to d(x,A)); |
| $U_\varepsilon(x)$ | $\underset{Df}{=} \{y \in R^n \,/\, d(x,y) < \varepsilon\}$, $\varepsilon$-neighbourhood of the point x; |
| $U_\varepsilon A$ | $\underset{Df}{=} \{y \in R^n \,/\, d(y,A) < \varepsilon\}$, $\varepsilon$-neighbourhood of the set A; |
| A x B | Cartesean product of the sets A and B; |
| A+B; A-B | set addition; set subtraction; |
| int A | interior of the set A; |
| cl A | closure of the set A; |
| aff A | affine hull of A; |
| dim A | dimension of A; |
| rel int A; ri A | interior of A relatively to aff A; |
| bd A | boundary of A |
| $A^P$ | $\underset{Df}{=} \{x \in R^n \,/\, x^T y \le 0,\ y \in A\}$; polar cone of A; |
| $A^\perp$ | $\underset{Df}{=} \{x \in R^n \,/\, x^T y = 0,\ y \in A\}$, orthogonal complement of the linear subspace A; |
| conv A | convex hull of A; |
| $0^+A$ | $\underset{Df}{=} \{u \in R^n \,/\, \exists\, x \in A:\ x + \lambda u \in A\ \forall\, \lambda \ge 0\}$, recession cone of the convex set A; |

The degree of a polynomial function $p : R^n \longrightarrow R$ is denoted by deg(p). A polynomial function p(x) with deg(p) = m is called a form of the order m if $p(tx) = t^m p(x)\ \forall\, x \in R^n,\ \forall\, t \in R$. Each polynomial function p(x) with $\deg(p) = m \ge 1$ possesses a representation by forms $q_j$, j=0,1,...,m,

$$p(x) = q_m(x) + \dots + q_1(x) + q_o$$

where deg $(q_j)$ is j or 0. A polynomial function is linear if $p(x) = q_1(x) + q_o$ and it is a polynomial without a linear term if $p(x) = q_m(x) + \dots + q_2(x)$.

The following letters have a fixed meaning throughout the whole text:

| | |
|---|---|
| M | constraint set-mapping (of a "continuous" optimization problem); |
| G | constraint set-mapping (of a (mixed-) integer problem); |
| $\varphi$ | value function; |
| $\psi$ | optimal set-function; |

N level set-mapping.

Finally, we use the abbreviations

(IG) due to Section 5.1.

(MIG) due to Section 5.1.

(CPC) due to Section 4.2.

for frequently imposed assumptions.

## 2. Multifunctions and Constraint Sets

This chapter contains some well-known basic results on the continuity properties of multifunctions (constraint sets) with images in the space $R^n$. The results presented here correspond to fairly general constraint sets which are closely related to applications in integer non-linear prametric programming. (Theorem 1. due to KLATTE and Theorem 3. and 4. originally du to KUMMER (1978)). More comprehensive and general investigations of multifunctions one can find e.g. in BANK et al. (1982).

### 2.1. Definitions

Let $\Lambda \subset R^m$ be a non-empty set. The notation multifunction is used for a (set-valued) function $\Gamma$ which assigns to each point $\lambda \in \Lambda$ a (possibly empty) subset $\Gamma(\lambda)$ of $R^n$. Thus a multifunction $\Gamma$ is here a mapping from $\Lambda$ into the power set $2^{R^n}$ of the space $R^n$.

The following continuity definitions will be used. A multifunction $\Gamma$ is said to be

(i) closed at a point $\lambda^o \in \Lambda$ if

$$\left.\begin{array}{ll} \forall \{\lambda^t\} \subset \Lambda & : \lambda^t \longrightarrow \lambda^o, \\ \forall \{x^t\} \subset R^n & : x^t \in \Gamma(\lambda^t),\ t = 1,\dots, \\ & \quad x^t \longrightarrow x^o \end{array}\right\} \Longrightarrow x^o \in \Gamma(\lambda^o);$$

(ii) upper semicontinuous (according to HAUSDORFF) at a point $\lambda^o \in \Lambda$ if

$$\forall \varepsilon > 0 \Longrightarrow \exists \delta = \delta(\varepsilon) > 0 : \quad \Gamma(\lambda) \subset U_\varepsilon \Gamma(\lambda^o)\ \forall \lambda \in U_\delta\{\lambda^o\};$$

(iii) lower semicontinuous (according to BERGE) at a point $\lambda^o \in \Lambda$ if

$$\left.\begin{array}{l} \forall \Omega \subset R^n,\ \Omega \text{ open}: \\ \Gamma(\lambda^o) \cap \Omega \neq \emptyset \end{array}\right\} \Longrightarrow \left\{\begin{array}{l} \exists \delta = \delta(\Omega) > 0: \\ \quad M(\lambda) \cap \Omega_o \neq \emptyset \\ \quad \forall \lambda \in U_\delta\{\lambda^o\}; \end{array}\right.$$

(iv) H-lower semicontinuous (according to HAUSDORFF) at a point $\lambda^o \in \Lambda$ if

$$\forall \varepsilon > 0 \Longrightarrow \exists \delta = \delta(\varepsilon) > 0: \quad \Gamma(\lambda^o) \subset U_\varepsilon \Gamma(\lambda)\ \forall \lambda \in U_\delta\{\lambda^o\};$$

(v) continuous at $\lambda^o$ if it is both upper and lower semicontinuous at $\lambda^o$;

(vi) H-continuous at $\lambda^o$ if it is upper and H-lower semicontinuous at $\lambda^o$.

For convenience we will use the following abbreviations throughout the text:

u.s.c. and l.s.c.

for upper semicontinuous and lower semicontinuous, respectively. Furthermore, we say that a multifunction is continuous (or u.s.c., or l.s.c., or H-l.s.c., or H-continuous) if it shows that respective property at every point $\lambda^o \in \Lambda$.

The properties of a multifunction to be locally Lipschitzian will be introduced later. For further (semi-) continuity-terminologies, connections and elementary properties we refer to BANK et al. (1982).

If two multifunctions $\Gamma_1 : \Lambda_1 \longrightarrow 2^{R^n}$ and $\Gamma_2 : \Lambda_2 \longrightarrow 2^{R^n}$ are given, then we define two composed multifunctions as follows:

$$(\Gamma_1 \cap \Gamma_2)(\lambda) \underset{\overline{D}f}{=} \Gamma_1(\lambda) \cap \Gamma_2(\lambda),$$
$$(\Gamma_1 + \Gamma_2)(\lambda) \underset{\overline{D}f}{=} \Gamma_1(\lambda) + \Gamma_2(\lambda);$$

their effective domains are given by

$$\Lambda \underset{\overline{D}f}{=} \{\lambda \in \Lambda_1 \cap \Lambda_2 / \Gamma_1(\lambda) \cap \Gamma_2(\lambda) \neq \emptyset\},$$
$$\Lambda \underset{\overline{D}f}{=} \{\lambda \in \Lambda_1 \cap \Lambda_2 / \Gamma_1(\lambda) + \Gamma_2(\lambda) \neq \emptyset\},$$

respectively.

## 2.2. Parameter Depending Constraint Sets

We shall start by stating some basic results on constraint sets of the following form:

$$M(\lambda) \underset{\overline{D}f}{=} \{x \in R^n / f_i(x, \lambda) \leq 0,\ i = 1,\dots,m\},\ \lambda \in \Lambda, \tag{1}$$

where we impose convexity or quasiconvexity requirements on the real-valued functions $f_i : R^n \times \Lambda \longrightarrow R$, $i = 1,\dots,m$, for every fixed $\lambda$. By M we denote the corresponding multifunction, which is called the *constraint set-mapping* and defined on

$$\Lambda = \operatorname{dom} M \underset{\overline{D}f}{=} \{\lambda \in R^q / M(\lambda) \neq \emptyset\}. \tag{2}$$

To show the u.s.c.-property of constraints sets to have the form (1) we apply the following lemma:

Lemma 1.

Let $\lambda^o \in \Lambda$, $S \subset R^n$ be a compact convex set and $x^o \in M(\lambda^o) \cap S$. Further, let the functions $f_i$, $i = 1,\dots,m$, be quasiconvex on $R^n$ for each $\lambda \in \Lambda$ and lower semicontinuous on $R^n \times \{\lambda^o\}$. Moreover, the functions $f_i(x^o, .)$, $i = 1,\dots,m$, are supposed to be upper semicontinuous at $\lambda^o$. Finally, for an $\varepsilon > 0$ and two sequences $\{\lambda^t\} \subset \Lambda$ and $\{x^t\} \subset R^n$ let $\lambda^t \longrightarrow \lambda^o$, $x^t \in M(\lambda^t)$, $x^t \notin \operatorname{cl} U_\varepsilon S$ $\forall t$. Then the intersection of the line segment $s(x^o, x^t)$ for an arbitrary t and the boundary $\operatorname{bd} U_\varepsilon S$ is a single point $z^t$, and the sequence $\{z^t\}$ has an accumulation point $z^*$ which belongs to the intersection $M(\lambda^o) \cap \operatorname{bd} U_\varepsilon S$.

Proof. Obviously, by the assumptions of this lemma the intersection $s(x^o, x^t) \cap \operatorname{bd} U_\varepsilon S$ contains, for every fixed t, a single point only and there exist numbers $\alpha_t \in (0,1)$ allowing the representation

$$z^t = \alpha_t x^t + (1-\alpha_t)x^o \in \operatorname{bd} U_\varepsilon S \quad \forall t. \tag{3}$$

The compactness of $\operatorname{bd} U_\varepsilon S$ without loss of generality * implies that

$$\exists z^* \in \operatorname{bd} U_\varepsilon S : z^t \longrightarrow z^*. \tag{4}$$

By the quasiconvexity of $f_i$ on $R^n$, for every $\lambda \in \Lambda$, it follows from (3) and $x^t \in M(\lambda^t)$ that

$$_i(z^t, \lambda^t) \le \max\{f_i(x^t, \lambda^t), f_i(x^o, \lambda^t)\} \le$$
$$\le \max\{0, f_i(x^o, \lambda^{t})\} .$$

The continuity conditions of $f_i$ and the assumptions on the sequences $\{\lambda^t\}$ and $\{x^t\}$ then according to (4) give

$$f_i(z^*, \lambda^o) \le \underline{\lim}\, f_i(z^t, \lambda^t) \le$$
$$\le \max\{0, \overline{\lim}\, f_i(x^o, \lambda^t)\} \le$$
$$\le \max\{0, f_i(x^o, \lambda^o)\} = 0$$

from which the proposition follows. #

Theorem 1.

**Let $\lambda^o \in \Lambda$ be and hold that**

(i) the functions $f_i$ are quasiconvex for all $\lambda \in \Lambda$ and all $i = 1,\dots,m$,

(ii) the functions $f_i$ are lower semicontinuous on $R^n \times \{\lambda^o\}$ for all $i = 1,\dots,m$,

(iii) $M(\lambda^o)$ is a bounded set.

**Then the multifunction M defined by (1) is u.s.c. at $\lambda^o$.**

Proof. If we can show that

$$\exists\, \delta > 0: M(\lambda) \subset \operatorname{cl} U_1 M(\lambda^o) \quad \forall \lambda \in U_\delta\{\lambda^o\} \tag{5}$$

holds, then one may verify the validity of Theorem 1. in the following way. Assume that M is not u.s.c. at $\lambda^o$, then there exist an $\varepsilon > 0$ and a sequence $\{\lambda^t\} \subset \Lambda$, $\lambda^t \longrightarrow \lambda^o$ with

$$M(\lambda^t) \setminus U_\varepsilon M(\lambda^o) \ne \emptyset \quad \forall t. \tag{6}$$

Let $x^t$, for all t, be elements of the difference sets (6). By (iii) and (5) the sequence $\{x^t\}$ has an accumulation point $x^*$ belonging to $M(\lambda^o)$, because of the lower semicontinuity property of the functions $f_i$ the closedness of M at $\lambda^o$ is guaranteed.

It follows that $U_\varepsilon M(\lambda^o)$ contains an infinite number of the points $x^t$ which contradicts (6).

Now, assume (5) is not valid. Then there exist for each t, a point $\lambda^t \in U_{1/t}\{\lambda^o\}$ and a point $x^t \in R^n$ satisfying

$$x^t \in M(\lambda^t) \text{ and } x^t \notin W \overset{Df}{=} \operatorname{cl} U_1 M(\lambda^o) \quad \forall t.$$

If we now set $S = M(\lambda^o)$, then the hypotheses of Lemma 1. are fulfilled Thus, there exist a point $z^* \in R^n$ and numbers $\alpha_t \in (0, 1)$ such that

$$z^* \in M(\lambda^o) \cap \operatorname{bd} W,$$
$$z^t = \alpha_t x^t + (1 - \alpha_t) x^o \in \operatorname{bd} W,$$

and without loss of generality*

$$z^t \longrightarrow z^*$$

holds. By definition of W, every point of the set $M(\lambda^0)$ must be an interior point of W and we arrive at a contradiction: the above assumption is false and (5) is valid. #

One has to remark that all of the hypotheses in Theorem 1. are essential for its validity. Even in the case: $f_i(x,\lambda) = p_i(x) - \lambda_i$, $i = 1,\dots,m$, where $p_i: R^n \longrightarrow R$ are convex polynomial functions one generally cannot omit the assumption (iii) (such an example due to SHIRONIN is given in BANK/BELOUSOV et al. (1986)).

Now, we assign to the constraint set $M(\lambda)$ the set of its Slater points

$$M^0(\lambda) \underset{Df}{=} \{x \in R^n \;/\; f_i(x,\lambda) < 0,\ i = 1,\dots,m\} \tag{7}$$

for every fixed $\lambda \in \Lambda$. The corresponding multifunction is denoted by $M^0$.

This mapping is of particular importance for the lower semicontinuity of the multifunction M defined by (1).

Lemma 2.

Let

(i) $\Gamma : \Lambda \longrightarrow 2^{R^n}$ be a multifunction l.s.c. at $\lambda^0$,

(ii) $f_i$, $i = 1,\dots,m$, be upper semicontinuous on $M^0(\lambda^0) \times \{\lambda^0\}$,

(iii) $(M \cap \Gamma)(\lambda^0) \subset \mathrm{cl}(M^0 \cap \Gamma)(\lambda^0)$.

Then the composed multifunction $(M \cap \Gamma)$ is l.s.c. at $\lambda^0$.

Proof.

If $\bar{x}$ is a point in $M^0(\lambda^0)$, then it follows from (ii) that there are, for each i, positive numbers $\varepsilon_i$ and $\delta_i$ such that

$$f_i(x,\lambda) < 0 \quad \forall (x,\lambda) \in U_{\varepsilon_i}\{\bar{x}\} \times U_{\delta_i}\{\lambda^0\}.$$

With $\bar{\varepsilon} = \min_i \varepsilon_i$ and $\bar{\delta} = \min_i \delta_i$ one obtains

$$U_{\bar{\varepsilon}}\{\bar{x}\} \subset M^0(\lambda) \quad \forall \lambda \in U_{\bar{\delta}}\{\lambda^0\}. \tag{8}$$

Let $\Omega$ be an open set satisfying

$$\Omega \cap (M^0 \cap \Gamma)(\lambda^0) \neq \emptyset$$

and, let $x^0$ be a point of this intersection. Then, for each $\varepsilon > 0$, there exists a point $x(\varepsilon) \in \mathrm{cl}(M^0 \cap \Gamma)(\lambda^0)$ with $d(x^0, x(\varepsilon)) < \varepsilon$. If $\varepsilon$ is sufficiently small, then

$$U_\varepsilon\{x(\varepsilon)\} \subset \Omega$$

and, taking (8) into account, $\exists\, \eta_1 > 0$: $U_\varepsilon\{x(\varepsilon)\} \subset M^0(\lambda)\ \forall \lambda \in U_{\eta_1}\{\lambda^0\}$ holds. By the l.s.c.-property of $\Gamma$ at $\lambda^0$ one also has

$$\exists\, \eta_2 > 0: \; U_\varepsilon\{x(\varepsilon)\} \cap \Gamma(\lambda) \neq \emptyset \quad \forall \lambda \in U_{\eta_2}\{\lambda^0\}.$$ Choosing now $\delta = \min(\eta_1, \eta_2)$ we obtain

$$\emptyset \neq \Gamma(\lambda) \cap M^0(\lambda) \cap U_\varepsilon\{x(\varepsilon)\} \subset$$
$$\subset (M \cap \Gamma)(\lambda) \cap U_\varepsilon\{x(\varepsilon)\} \qquad \forall \lambda \in U_\delta\{\lambda^0\},$$

which completes the proof. #

Theorem 2.

Let the following hypotheses be satisfied:

(i) the multifunction $\Gamma$ is l.s.c. at $\lambda^o$,

(ii) $\Gamma(\lambda^o)$ is a convex set and $(M^o \cap \Gamma)(\lambda^o) \neq \emptyset$,

(iii) the functions $f_i$, $i = 1,\dots,m$, are upper semicontinuous on $M^o(\lambda^o) \times \{\lambda^o\}$,

(iv) the functions $f_i(.,\lambda^o)$, $i = 1,\dots,m$, are strictly quasiconvex on $\Gamma(\lambda^o)$.

Then the composed multifunction $(M \cap \Gamma)$ is l.s.c. at $\lambda^o$.

Proof.

Taking Lemma 2. into account we have only to show that

$$(M \cap \Gamma)(\lambda^o) \subset \mathrm{cl}(M^o \cap \Gamma)(\lambda^o)$$

holds. By assumption, there is a point $\bar{x} \in (M^o \cap \Gamma)(\lambda^o)$. For an arbitrarily fixed point $x \in (M \cap \Gamma)(\lambda^o)$ we define

$$x(\alpha) = \alpha\bar{x} + (1-\alpha)x, \quad \alpha \in (0,1).$$

Since the strictly quasiconvex function $f_i(.,\lambda^o)$ (i arbitrarily fixed) shows the property: if for $z,y \in \Gamma(\lambda^o)$, $z \neq y$, it holds that $f_i(z,\lambda^o) = f_i(y,\lambda^o)$, then there is at most one $v \in \mathrm{ri}\ s(z,y)$ such that $f_i(v,\lambda^o) > f_i(z,\lambda^o)$, there must be an $\alpha_i > 0$ such that

$$f_i(x(\alpha),\lambda^o) \leq f_i(\bar{x},\lambda^o) < 0 \quad \forall \alpha \in (0,\alpha_i),$$

if $f_i(x,\lambda^o) = f_i(\bar{x},\lambda^o)$. On the other hand, if $f_i(x,\lambda^o) \neq f_i(\bar{x},\lambda^o)$ holds, then one obtains by the definition of a strictly quasiconvex function that

$$f_i(x(\alpha),\lambda^o) < \max\{f_i(\bar{x},\lambda^o),\ f_i(x,\lambda^o)\} \leq 0 \quad \forall \alpha \in (0,1).$$

Setting $\alpha_i = 1$ in the latter case we conclude

$$f_i(x(\alpha),\lambda^o) < 0,\quad i=1,\dots,m,\quad \forall \alpha \in (0,\min_i \alpha_i).$$

Now, the convexity of $\Gamma(\lambda^o)$ implies that

$$x(\alpha) \in M^o(\lambda^o) \cap \Gamma(\lambda^o) \quad \forall \alpha \in (0,\min_i \alpha_i)$$

holds. Taking $\alpha \longrightarrow 0$ one obtains the validity of this theorem. #

If we consider the particular case

$$f_i(x,\lambda) = g_i(x) - \lambda_i,\quad i = 1,\dots,m,$$

we obtain the constraint sets in one of our standard forms

$$M(\lambda) \underset{\mathrm{Df}}{=} \{x \in R^n \ /\ g_i(x) \leq \lambda_i,\ i = 1,\dots,m\}\ . \qquad (10)$$

Theorem 3.

Let the functions $g_i$, $i = 1,\dots,m$, be convex and, let the set $M(\lambda^o)$ defined by (10) be a non-empty affine subspace. Then the multifunction M is continuous at $\lambda^o$.

Proof. There is a linear subspace U such that

$$M(\lambda^o) = x^o + U$$

holds for every $x^o \in M(\lambda^o)$. By the convexity of the functions $g_i$ $i = 1,\dots,m$, one knows that

$$f_i(x + \alpha u) = f_i(x) \quad \forall i \quad \forall \alpha \in R \quad \forall x \in R^n \; \forall u \in U \tag{9}$$

must be valid.

If dim U = 0 (i.e. $U = \{0\}$ ), then there is nothing to prove. Assume that the proposition is already shown for the case dim U < k. We consider the case that dim U = k. For a fixed point $v \in U$, $v \neq 0$, let us define

$$M_v(\lambda) = \{x \in M(\lambda) \;/\; v^T x = 0\}.$$

Then, by (9), it follows that

$$M(\lambda) = M_v(\lambda) + L_v$$

may be written where

$$L_v = \{u \in R^n \;/\; u = \alpha v, \alpha \in R\}.$$

$M_v(\lambda)$ is a linear subspace with dim $M_v(\lambda) < k$ for which the proposition is true. Therefore, according to $M(\lambda) = M_v(\lambda) + L_v$ the proposition is also true for the multifunction M. #

A convex function $f: R^n \longrightarrow R$ is called weakly analytic if the following implication holds: when f is constant on a line segment S, then it is constant on the whole line generated by S.

Theorem 4.

Let all the functions $g_i$, $i = 1,\dots,m$, be convex and $\lambda^o \in \Lambda$ be fixed. If there is a point $x^o \in M(\lambda^o)$ such that $g_i(x^o) < \lambda_i^o$, for every i for which $g_i$ is not weakly analytic, then the multifunction M defined by (10) is l.s.c. at $\lambda^o$.

Proof. We define the following index sets

$$J \overset{=}{\text{Df}} \{i \in \{1,\dots,m\} / \; g_i \text{ is weakly analytic}\},$$

$$I_1 \overset{=}{\text{Df}} \{i \in \{1,\dots,m\} / \; g_i(x) = \lambda_i^o \;\forall x \in M(\lambda^o)\},$$

$$I_2 \overset{=}{\text{Df}} \{i \in \{1,\dots,m\} / \; \exists x(i) \in M(\lambda^o): \; g_i(x(i)) < \lambda_i^o\}.$$

Then $I_1 \subset J$ holds and, the definition of a weakly analytic function implies that

$$M^{I_1}(\lambda^o) \underset{Df}{=} \{x \in R^n \ / \ g_i(x) \leq \lambda_i^o \quad \forall i \in I_1\} \tag{11}$$

is an affine subspace which coincides with the space $R^n$ if $I_1 = \emptyset$.
According to Theorem 3. the multifunction $M^{I_1}: \Lambda \longrightarrow 2^{R^n}$ defined by (11) is continuous at $\lambda^o$.
Defining now the set

$$M^{I_2}(\lambda^o) \underset{Df}{=} \{x \in R^n \ / \ g_i(x) < \lambda_i^o \quad \forall i \in I_2\} \tag{12}$$

(for which we put $M^{I_2}(\lambda^o) \underset{Df}{=} R^n$ If $I_2 = \emptyset$)

one easily confirms that

$$M(\lambda^o) = cl(M^{I_1}(\lambda^o) \cap M^{I_2}(\lambda^o)) \tag{13}$$

and

$$M(\lambda) \supset cl(M^{I_1}(\lambda) \cap M^{I_2}(\lambda)) \quad \forall \lambda \in \Lambda \tag{14}$$

hold. In order to prove that M is l.s.c. at $\lambda^o$ we consider an arbitrary sequence $\{\lambda^t\} \subset \Lambda$, $\lambda^t \longrightarrow \lambda^o$ and a point $\bar{x} \in M^{I_1}(\lambda^o) \cap M^{I_2}(\lambda^o)$.
According to the continuity of $M^{I_1}$ at $\lambda^o$ there is a sequence $\{x^t\} \subset R^n$ with the properties $x^t \in M^{I_1}(\lambda^t)$ and $x^t \longrightarrow \bar{x}$.
Since $\bar{x}$ belongs to $M^{I_2}(\lambda^o)$ one has by (12) for a sufficiently small $\varepsilon > 0$,

$$U_\varepsilon(\bar{x}) \subset M^{I_2}(\lambda^o).$$

Hence, it follows that

$$x^t \in M^{I_2}(\lambda^o) \text{ for almost all } t.$$

Therefore, the multifunction $(M^{I_1} \cap M^{I_2})$ is l.s.c. at $\lambda^o$ and, taking (13) and (14) into account, the same holds for M. #

Now we introduce multifunctions $M: \Lambda \longrightarrow 2^{R^n}$ defined by

$$M(\lambda) \underset{Df}{=} C(\lambda) + V, \quad \lambda \in \Lambda \subset R^m, \tag{15}$$

where $C: \Lambda \longrightarrow 2^{R^n}$ is an u.s.c. compact valued multifunction having non-empty images and V is a polyhedral cone.

The following theorem shows that the class of multifunctions defined by (15) is closed with respect to the composition of finitely many such multifunctions by intersection. The same is true for the composition of the form $(M_1+M_2)$, which one easily confirms.

<u>Theorem 5.</u>
Let $C_1 : \Lambda_1 \longrightarrow 2^{R^n}$ and $C_2: \Lambda_2 \longrightarrow 2^{R^n}$ be two compact-valued u.s.c. multifunctions defined on $\Lambda_1, \Lambda_2 \subset R^m$ and $V_1, V_2 \subset R^n$ polyhedral cones.
Further, let

$$V \underset{Df}{=} V_1 \cap V_2,\quad M(\lambda) \underset{Df}{=} \bigcap_{i=1}^{2} (C_i(\lambda)+V_i),$$

$$\Lambda \underset{Df}{=} \{\lambda \in R^m \;/\; M(\lambda) \neq \emptyset\}.$$

Then for the multifunction $M : \Lambda \longrightarrow 2^{R^n}$ defined by $M(\lambda)$ there exists a compact-valued u.s.c. multifunction $C: \Lambda \longrightarrow 2^{R^n}$ such that

$$M(\lambda) = C(\lambda) + V \quad \forall \lambda \in \Lambda \tag{16}$$

holds. Further, M itself is an u.s.c. multifunction.

Proof.

Let H be the subspace of maximal dimension contained in the polyhedral cone V. By $H^\perp$ we denote the orthogonal complement of H. Further, let $S^1,\dots,S^q$ and $T^1,\dots,T^p$ be the closed faces of $V_1 \cap H^\perp$ and $V_2 \cap H^\perp$, respectively.

We show that the multifunction C defined by

$$C(\lambda) \underset{Df}{=} \bigcup_{(i,j):S^i \cap T^j = \{0\}} \{(C_1(\lambda)+S^i) \cap (C_2(\lambda)+T^j)\}$$

has the required properties.

First, we show (16). Obviously, $M(\lambda) \supset C(\lambda)+V$. For an arbitrary $\lambda \in \Lambda$ let $x \in M(\lambda)$ be an arbitrary point. We select now closed faces $S^i$ and $T^j$ of $V_1 \cap H^\perp$ and $V_2 \cap H^\perp$, respectively, which are of smallest dimension fulfilling

$$x \in [(C_1(\lambda)+S^i)+V] \cap [(C_2(\lambda)+T^j)+V].$$

Then there are points

$$c^l \in C_l(\lambda),\; v^l \in V,\; l = 1,2,\; u^1 \in S^i \text{ and } u^2 \in T^j$$

such that

$$x = c^l + u^l + v^l \text{ for } l = 1,2$$

hold. Because of the minimality of the dimensions of $S^i$ and $T^j$ we have

$$u^1 \in \text{ri } S^i \text{ and } u^2 \in \text{ri } T^j.$$

In order to prove that $x \in C(\lambda)+V$, it suffices to show that $S^i \cap T^j = \{0\}$. Assuming

$$\exists u \neq 0 : u \in S^i \cap T^j$$

for $\varepsilon > 0$ and $\varepsilon$ sufficiently small we have

$$u^1 - \varepsilon u \in S^i \text{ and } u^2 - \varepsilon u \in T^j.$$

Since $\{0\}$ is the only linear subspace contained in the intersection $(V_1 \cap H^\perp) \cap (V_2 \cap H^\perp)$, there exists an $\varepsilon^* > 0$ such that

$$u^1 - \varepsilon^* u \in S^{i^*} \text{ and } u^2 - \varepsilon^* u \in T^{j^*}$$

where $S^{i^*}$ and $T^{j^*}$ are closed faces of $(V_1 \cap H^\perp)$ and $(V_2 \cap H^\perp)$, respectively, and it holds

$$\dim S^{i*} < \dim S^i \text{ or } \dim T^{j*} < \dim T^j.$$

Taking into account that, for $l = 1,2$,

$$v^l + \varepsilon^* u \in V,$$

$$x = c^l + (u^l - \varepsilon^* u) + (v^l + \varepsilon^* u)$$

hold one obtains a contradiction to the selection of $S^i$ or $T^j$. Hence, $S^i \cap T^j = \{0\}$ must be valid.

Obviously, $C(\lambda)$ is a compact set.

Let $\lambda^0 \in \Lambda$ be an arbitrary point. The u.s.c. property of the multifunctions $C_1$ and $C_2$ implies the existence of a compact set A and a neighbourhood $U_\delta(\lambda^0)$ such that

$$C(\lambda) \subset A \quad \forall \lambda \in U_\delta(\lambda^0).$$

The closedness of C at $\lambda^0$ yields now that C is u.s.c. at $\lambda^0$ and by (16) the same is true for M. #

## 3. Stability of Some Continuous Optimization Problems

In this chapter we give a survey on stability properties of parametric optimization problems in the space $R^n$, i.e. we derive continuity properties of the value function and the optimal set-function. Thereby certain convexity requirements permit to state the results without imposing compactness on the involved sets. Theorems 1. and 2. are valid in normed spaces (e.g. see BANK et al. (1982)). The restriction to $R^n$ made here corresponds to the aim of this monograph. In the Theorems 1., 2. and 3. an explicit representation of the constraint sets is not underlaid; a representation of the convex constraint sets by means of real-valued functions is used in the Theorems 4. and 5. The Theorems 6. and 7. are devoted to parametric problems with fixed constraint sets which have only to be closed and, therefore the results may directly be applied to mixed-integer problems with closed constraint sets. **The Theorems 1. to 6. can be found in BANK et al. (1982) (1., 2. and 5. originally due to KUMMER (1978), 3., 4. and 6. to KLATTE (1977)). The proof of Theorem 7. bases on an idea of ANDRONOW/BELOUSOV (1985).**

It should be noted that the assumptions imposed in this chapter are not just of technical nature for establishing the proofs. If one omits any of them, then the statements are no longer valid; for counter-examples the reader is referred to BANK et al. (1982).
Further stability investigations of continuous parametric optimization problems involving only (quasi)convex polynomial functions are executed below in Section 4.3. Finally, the results for (mixed-)integer problems in Chapter 6. include the case in which no integer requirements are imposed on the variables.

### 3.1. Definitions

Let a parametric optimization problem be defined by

$$P(\lambda) \quad \inf \{f(x,\lambda) \;/\; x \in M(\lambda)\} \;, \lambda \in \Lambda \subset R^m,$$

which we will understand as follows: for every arbitrarily fixed point $\lambda \in \Lambda$ the real valued objective function $f(.,\lambda)$ has to be minimized subject to all points x belonging to the fixed set $M(\lambda)$. The constraint sets $M(\lambda)$ are given by a multifunction $M : \Lambda \longrightarrow 2^{R^n}$ as considered in Chapter 2. Obviously, for a general study of $P(\lambda)$ it is not necessary to distinguish between components of $\lambda$ appearing only in the objective function and such appearing only in the constraint set $M(\lambda)$, respectively.
Since in the following we mainly consider the behaviour of $P(\lambda)$ in terms of the extreme value and the optimal set as functions on $\Lambda$; for simplicity, we restrict ourselves to the case that

$$\Lambda \subset \text{dom } M.$$

The *value function* $\varphi : \Lambda \longrightarrow R$ is defined by

$$\varphi(\lambda) \underset{Df}{=} \inf \{f(x,\lambda) \mid x \in M(\lambda)\} .$$

Further, the multifunction $\Psi : \Lambda \longrightarrow 2^{R^n}$ given by the optimal sets

$$\Psi(\lambda) \underset{Df}{=} \{x \in M(\lambda) \mid f(x,\lambda) = \varphi(\lambda)\},$$

is called the *optimal set-function*. Its domain dom $\Psi$ is the *solvability set* of the parametric optimization problem $P(\lambda)$. As usual in parametric optimization, stability properties of $P(\lambda)$ are characterized by semi-continuity properties of the mappings $M$, $\varphi$ and $\Psi$ associated to $P(\lambda)$. In the literature a unified notation of stability of optimization problems does not exist, here we use the notation "stability"as an abbreviation for certain continuity properties of the mappings above.

## 3.2. Stability Properties

Theorem 1.

Let the following assumptions hold:

(i) $T \subset R^n$ and $\Omega \subset R^m$ are open convex sets,

(ii) $\Gamma : \Omega \longrightarrow 2^{R^n}$ is a multifunction which is l.s.c. at $\lambda^0 \in \Omega$ and fulfils the condition

$$\tfrac{1}{2}\Gamma(\lambda^1) + \tfrac{1}{2}\Gamma(\lambda^2) \subset \Gamma(\tfrac{1}{2}\lambda^1 + \tfrac{1}{2}\lambda^2) \quad \forall \lambda^1, \lambda^2 \in \Omega ,$$

(iii) $f$ is independent of $\lambda$ and convex on the set $T$,

(iv) $M$ is defined by

$$M(\lambda) \underset{Df}{=} \Gamma(\lambda) \cap T \qquad \forall \lambda \in \Omega,$$

(v) $M(\lambda^0) \neq \emptyset$.

Then the value function $\varphi$ is continuous at $\lambda^0$.

Proof.

First, we show that the multifunction $M$ is l.s.c. at $\lambda^0$. Let $W$ be an open set with $M(\lambda^0) \cap W \neq \emptyset$ and $x$ a point belonging to this intersection. Then by definition of $M$ the following holds:

$$\forall \varepsilon > 0 \ \exists y^\varepsilon \in M(\lambda^0) : \ \|x - y^\varepsilon\| < \varepsilon$$

and, if $\varepsilon$ is taken sufficiently small, the point $y^\varepsilon$ belongs to $W$. Furthermore there is a neighbourhood $W' \subset W$ of $y^\varepsilon$ because $T$ is an open set. The lower semicontinuity of $\Gamma$ at $\lambda^0$ ensures the existence of a $\delta > 0$ such that

$$\Gamma(\lambda) \cap W' \neq \emptyset \quad \forall \lambda \in U_\delta\{\lambda^0\}.$$

Hence, the multifunction $M$ is l.s.c. at $\lambda^0$.

Let $\varepsilon > 0$ and $x^0 \in M(\lambda^0)$ such that $f(x^0) < \varphi(\lambda^0) + \varepsilon$. Then we know by the l.s.c.-property of $M$ at $\lambda^0$ that for each sequence $\{\lambda^t\} \subset \Omega$ with $\lambda^t \longrightarrow \lambda^0$ there exists a sequence $\{x^t\} \subset R^n$ such that

$$x^t \in M(\lambda^t) \text{ for almost all } t \text{ and } x^t \longrightarrow x^0.$$

Using the continuity of f over T one obtains

$$\overline{\lim}\, \varphi(\lambda^t) \le \overline{\lim}\, f(x^t) \le f(x^0) < \varphi(\lambda^0) + \varepsilon .$$

Hence, the value function is upper semicontinuous at $\lambda^0$.

In order to show that $\varphi$ is also lower semicontinuous at $\lambda^0$ we choose an arbitrary $\varepsilon > 0$. By the property just proven for $\varphi$ there is a $\delta_0 > 0$ such that

$$\varphi(\lambda) < \varphi(\lambda^0) + \varepsilon \quad \forall \lambda \in U_{\delta_0}\{\lambda^0\}. \tag{1}$$

If $\delta_0$ is chosen sufficiently small, then $U_\delta\{\lambda^0\} \subset \Omega$ and, obviously, for a fixed pair of points $\lambda \in U_{\delta_0}\{\lambda^0\}$ and $x \in M(\lambda)$ the point $\mu = \lambda^0 + (\lambda^0 - \lambda)$ also belongs to $U_{\delta_0}\{\lambda^0\}$ (here, the case $M(\lambda) = \emptyset$ is trivial). According to (1) and $\varphi(\lambda^0) < \infty$ there exists a $z \in M(\mu)$ satisfying $f(z) < \varphi(\lambda^0) + \varepsilon$.

By the hypotheses (i) and (ii) one has

$$1/2x + 1/2z \in M(\lambda^0),$$

from which one obtains, employing the convexity of f, that

$$\varphi(\lambda^0) \le f(1/2x + 1/2z) < 1/2f(x) + 1/2(\varphi(\lambda^0) + \varepsilon),$$

i.e.

$$f(x) > \varphi(\lambda^0) - \varepsilon.$$

But this shows, by the choice of $\varepsilon$ and x, the existence of a $\delta > 0$ such that

$$\varphi(\lambda) \ge \varphi(\lambda^0) - \varepsilon \quad \forall \lambda \in U_\delta\{\lambda^0\} . \quad \#$$

Theorem 2.

The value function $\varphi$ is lower semicontinuous at $\lambda^0$ if the following hypotheses are satisfied:

(i) M is u.s.c. at $\lambda^0 \in \Lambda$

(ii) $M(\lambda^0)$ is a convex set,

(iii) f is independent of $\lambda$ and convex.

Proof.

If one defines a multifunction $\Gamma$ by

$$\Gamma(z) \underset{Df}{=} z + M(\lambda^0),$$

then the assumption (ii) of the Theorem 1. is fulfilled. Then the value function

$$h(z) = \inf \{f(x) \,/\, x \in \Gamma(z)\} , \quad z \in R^n$$

is continuous over $R^n$ according to the same theorem.

Now, let $\{\lambda^t\} \subset \Lambda$ be an arbitrary sequence with $\lambda^t \longrightarrow \lambda^0$. Since

M is u.s.c. at $\lambda^0$ one concludes for large t that

$$\exists \varepsilon_t > 0 : M(\lambda^t) \subset \bigcup_{\|z\|<\varepsilon_t} \Gamma(z), \quad \varepsilon_t \longrightarrow 0.$$

Thus

$$\underline{\lim}\, \varphi(\lambda^t) \geq \underline{\lim}\, \inf \{h(z) \,/\, \|z\| < \varepsilon_t\} =$$
$$= h(0) = \varphi(\lambda^0). \qquad \#$$

For the next two theorems we will impose the following hypotheses:

(2) the constraint set mapping M is lower semicontinuous at $\lambda^0$,

(3) the optimal set $\Psi(\lambda^0)$ is non-empty and bounded,

(4) the objective function f is lower semicontinuous on $R^n \times \{\lambda^0\}$,

(5) there exists a point $x^0 \in \Psi(\lambda^0)$ such that $f(x^0,.)$ is upper semi-continuous at $\lambda^0$,

(6) $f(.,\lambda)$ is a quasiconvex function for each fixed $\lambda \in \Lambda$,

(7) the constraint set $M(\lambda^0)$ is a closed convex set.

Theorem 3.

Suppose that

(i) the hypotheses (2) - (7) hold,

(ii) the constraint set $M(\lambda)$ is convex for every $\lambda \in \Lambda$,

(iii) the constraint set mapping M is closed at $\lambda^0$.

Then the value function $\varphi$ is continuous at $\lambda^0$ and the optimal set-function $\Psi$ is u.s.c. at $\lambda^0$.

Proof.

First we shall show that $\varphi$ is upper semicontinuous at $\lambda^0$. According to (5) there is a point $x^0 \in \Psi(\lambda^0)$ such that the objective function f is upper semicontinuous at $(x^0, \lambda^0)$. Then by (2) there exist points $x^t \in M(\lambda^t)$ such that $x^t \longrightarrow x^0$ for each arbitrarily chosen sequence $\{\lambda^t\} \subset \Lambda$ with $\lambda^t \longrightarrow \lambda^0$ and t sufficiently large. Thus, it follows

$$\overline{\lim}\, \varphi(\lambda^t) \leq \overline{\lim}\, f(x^t, \lambda^t) \leq f(x^0, \lambda^0) = \varphi(\lambda^0).$$

The next step will be to prove that the optimal set-function $\Psi$ is u.s.c. at $\lambda^0$. We define the multifunction $N : \Lambda \longrightarrow 2^{R^n}$ by

$$N(\lambda) \underset{Df}{=} \{x \in R^n \,/\, f(x,\lambda) - \varphi(\lambda) \leq 0\}.$$

Obviously, then we have

$$\Psi(\lambda) = M(\lambda) \cap N(\lambda) \qquad \forall \lambda \in \Lambda.$$

Furthermore,

$$W \underset{Df}{=} \operatorname{cl} U_1(M(\lambda^0) \cap N(\lambda^0)).$$

If one now can show that

$$\exists \delta > 0 : \Psi(\lambda) \subset W \qquad \forall \lambda \in U_\delta\{\lambda^0\}, \tag{8}$$

then the upper semicontinuity of $\Psi$ is obtained as follows.

The multifunction N is closed at $\lambda^0$ because for every sequence $\{\lambda^t\} \subset \Lambda$ with $\lambda^t \longrightarrow \lambda^0$ and for every sequence $\{x^t\} \subset R^n$ having the properties $x^t \in N(\lambda^t)$ and $x^t \longrightarrow x^*$ one obtains

$$f(x^*, \lambda^0) - \varphi(\lambda^0) \leq \underline{\lim}(f(x^t, \lambda^t) - \varphi(\lambda^t)) \leq 0$$

since f is lower semicontinuous on $R^n \times \{\lambda^0\}$ and $-\varphi$ is lower semicontinuous at $\lambda^0$; hence $x^* \in N(\lambda^0)$.

Then by (iii), the optimal set-function $\Psi$ is closed at $\lambda^0$, too.

Further, for every pair of sequences $\{\lambda^t\} \subset \Lambda$, $\{x^t\} \subset R^n$ such that $\lambda^t \longrightarrow \lambda^0$ and $x^t \in \Psi(\lambda^t)$ there exists an accumulation point $x^*$ of $\{x^t\}$ with $x^* \in \Psi(\lambda^0)$. This is true because of the hypothesis (3), the closedness of $\Psi$ at $\lambda^0$ and (8). Now, the upper semicontinuity of $\Psi$ at $\lambda^0$ is an immediate consequence.

It remains to verify (8), which we carry out by contradiction. If (8) is false, then there exist, for every t = 1,2,..., points $\lambda^t$ and $x^t$:

$$\lambda^t \in U_{1/t}\{\lambda^0\}, \; x^t \in \Psi(\lambda^t), \; x^t \notin W.$$

Considering the interior point $x^0$ of W existing by (5), we construct the sequence $\{z^t\}$ of all intersection points between the line segments $s(x^0, x^t)$ and the boundary bd W of W. We show that this sequence $\{z^t\}$ has an accumulation point

$$z^* \in \Psi(\lambda^0) \cap \text{bd } W. \tag{9}$$

Obviously, every accumulation point of $\{z^t\}$ belongs to bd W. In order to verify the existence of a desired accumulation point $z^*$ we use the representation

$$z^t = \alpha_t x^t + (1-\alpha_t)x^0, \quad \alpha_t \in (0,1) \; \forall t \tag{10}$$

for the members of the sequence and, without loss of generality * we underly that

$$z^t \longrightarrow z^*.$$

Employing the lower semicontinuity properties of f and $-\varphi$ and the quasiconvexity of f one concludes

$$f(z^*, \lambda^0) - \varphi(\lambda^0) \leq \underline{\lim}(f(z^t, \lambda^t) - \varphi(\lambda^t)) \leq$$
$$\leq \max(\underline{\lim}(f(x^t, \lambda^t) - \varphi(\lambda^t)), \overline{\lim}(f(x^0, \lambda^t) - \varphi(\lambda^t))).$$

Now, since both $f(x^t, \lambda^t) = \varphi(\lambda^t)$ for each t and the upper semicontinuity of f at $(x^0, \lambda^0)$ hold, the right-hand side of the above inequality must be less than or equal to zero, thus $z^* \in N(\lambda^0)$. For the validity of (9) it remains to show that $z^*$ is a point of $M(\lambda^0)$. By (2) it follows that there exists a sequence $\{y^t\}$ such that $y^t \in M(\lambda^t)$ and $y^t \longrightarrow x^0$. Using now the real numbers $\alpha_t \in (0,1)$, for all t, from (10) we define points

$$\bar{y}^t = \alpha_t x^t + (1-\alpha_t)y^t$$

which by (ii) are in $M(\lambda^t)$.
From (10) we have

$$\bar{y}^t = z^t + (1-\alpha_t)(y^t - x^o) \quad \forall t,$$

and since both $y^k \longrightarrow x^o$ and $z^t \longrightarrow z^*$, we conclude that $\bar{y}^t \longrightarrow z^*$, Now, (iii) ensures $z^* \in M(\lambda^o)$, which completes the proof of (9). But, (9) leads to a contradiction to the definition of W and, therefore, our assumption that (8) is not true was false.

We still have to demonstrate that the value function $\varphi$ is lower semicontinuous at $\lambda^o$. Therefore, we suppose that $\varphi$ is not lower semicontinuous at $\lambda^o$. Then there exist a sequence $\{\lambda^t\}$ with $\lambda^t \longrightarrow \lambda^o$ and an $\varepsilon > 0$ such that

$$\varphi(\lambda^t) < \varphi(\lambda^o) - \varepsilon. \tag{11}$$

Let, for each t, $x^t \in M(\lambda^t)$ be an arbitrary point with $f(x^t, \lambda^t) < \varphi(\lambda^o) - \frac{\varepsilon}{2}$, then it follows that

$$\overline{\lim} f(x^t, \lambda^t) < \varphi(\lambda^o) \tag{12}$$

holds.

Defining

$$G(\lambda) \underset{Df}{=} \{x \in R^n \,/\, f(x,\lambda) \le \varphi(\lambda^o)\} \cap M(\lambda^o)$$

we obtain from (12) that $x^t \in G(\lambda^t)$ and, that the function h given by

$$h(\lambda) \underset{Df}{=} \inf \{f(x,\lambda) \,/\, x \in G(\lambda)\}$$

cannot be lower semicontinuous at $\lambda^o$.

On the other hand, following a similar way as used above for the multifunction $\Psi$ one may show that G is u.s.c. at $\lambda^o$ (hint: $G(\lambda^o) \equiv \Psi(\lambda^o)$ is a compact set). Hence, the sequence $\{x^t\}$ has an accumulation point $x^* \in G(\lambda^o)$ and without loss of generality * we assume that $x^t \longrightarrow x^*$. Now, the lower semicontinuity of f at $(x^*, \lambda^o)$ implies that

$$\overline{\lim}\, f(x^t, \lambda^t) \ge \underline{\lim}\, f(x^t, \lambda^t) \ge$$
$$\ge f(x^*, \lambda^o) \ge h(\lambda^o)$$

holds, contradicting the above conclusion of (12) that h cannot be lower semicontinuous at $\lambda^o$.

This completes the proof of Theorem 3. #

Remark 1.

The preceding Theorem 3. allows the following two consequences for the so-called $\varepsilon$-optimal sets. If the hypotheses of Theorem 3. are fulfilled, then

(i) the $\varepsilon$-optimal set mapping $\Psi_\varepsilon : \Lambda \longrightarrow 2^{R^n}$ given by

$$\Psi_\varepsilon(\lambda) \underset{Df}{=} \{x \in M(\lambda) \,/\, f(x,\lambda) < \varphi(\lambda) + \varepsilon\},\ \varepsilon > 0,$$

is l.s.c. at $\lambda^o$;

(ii) the mapping $\overline{\Psi} : \Lambda \times R_+ \longrightarrow 2^{R^n}$ defined by

$$\bar{\Psi}(\lambda,\varepsilon) \underset{Df}{=} \{x\in M(\lambda) \ / \ f(x,\lambda) \le \varphi(\lambda) + \varepsilon\}$$

is u.s.c. at the point $(\lambda^o, 0)$.

The proofs are carried out in BANK et al. (1982).

Theorem 4.

Let the constraint set $M(\lambda)$ of $P(\lambda)$ be given in the form (2.1). Further, let the assumption (3) hold and let (4), (5) and (6) be required for both the objective function f and the functions $f_i$, $i=1,\dots,m$, which appear in the constraints. Then the value function $\varphi$ is lower semicontinuous at $\lambda^o$.

Proof.

As in the proof of Theorem 3, we define

$$G(\lambda) \underset{Df}{=} M(\lambda) \cap \{x\in R^n \ / \ f(x,\lambda) \le \varphi(\lambda^o)\} =$$

$$= \{x\in R^n \ / \ g_i(x,\lambda) \le 0,\ i=1,\dots,m; \quad f(x,\lambda) - \varphi(\lambda^o) \le 0\},\ \lambda\in\Lambda .$$

Obviously, $G(\lambda^o) \equiv \Psi(\lambda^o)$ and therefore, by our assumptions, we may apply the Theorem 2.1., from which we can conclude that the multifunction G defined above is u.s.c. at $\lambda^o$.

Thus the value function

$$h(\lambda) \underset{Df}{=} \inf\{f(x,\lambda) \ / \ x\in G(\lambda)\}$$

is lower semicontinuous at $\lambda^o$, which one analogously may verify as in the proof of the preceding theorem. Therefore, also the value function $\varphi$ must show this property. #

The next theorem is related to the more specialized parametric problem

$$P(\lambda) \quad \inf\{f(x) \ / \ g_i(x) \le \lambda_i,\ i=1,\dots,m\},\ \lambda\in\Lambda,$$

i.e. the objective function is independent of $\lambda$ and the constraint set $M(\lambda)$ is given in the form (2.10).

Theorem 5.

Let both the objective function f and the constraint functions $g_i$, $i=1,\dots,m$, be convex. Further, let

$$J \underset{Df}{=} \{i \ / \ g_i \text{ weakly analytic}\}$$

be defined and $\lambda^o$ a point in $\Lambda \subset \operatorname{dom} M$. Then the following statements hold:

(i) If there is a point $x^o\in M(\lambda^o)$ such that $g_i(x^o) < \lambda_i^o$ for all $i\notin J$, then the value function $\varphi$ of $P(\lambda)$ is continuous at $\lambda^o$ and the corresponding optimal set-function $\Psi$ is closed at $\lambda^o$.

(ii) If the optimal set $\Psi(\lambda)$ is nonempty for all $\lambda \in \Lambda$ and all the functions f and $g_i$, i=1,...,m, are weakly analytic, then the optimal set-function $\Psi$ is l.s.c. at $\lambda^0$.

Proof.
(i) At first we show that $\varphi$ is upper semicontinuous at $\lambda^0$. Let $\varepsilon > 0$ and $x^0 \in M(\lambda^0)$ be such that

$$f(x^0) < \varphi(\lambda^0) + \varepsilon .$$

According to Theorem 2.4. the multifunction M is l.s.c. at $\lambda^0$. Therefore,

$$\forall \{\lambda^t\} \subset \Lambda : \lambda^t \longrightarrow \lambda^0 \; \exists \{x^t\} \subset R^n : x^t \in M(\lambda^t),\; x^t \longrightarrow x^0 .$$

Hence, using the upper semicontinuity of f at $\lambda^0$ one concludes that

$$\overline{\lim}\, \varphi(\lambda^t) \leq \overline{\lim}\, f(x^t) \leq f(x^0) < \varphi(\lambda^0) + \varepsilon ,$$

which implies that $\varphi$ is upper semicontinuous at $\lambda^0$.
In order to show that $\Psi$ is closed at $\lambda^0$ we consider arbitrary sequences $\{\lambda^t\} \subset \Lambda$ and $\{x^t\} \subset R^n$ with $\lambda^t \longrightarrow \lambda^0$ and $x^t \in \Psi(\lambda^t)$, $x^t \longrightarrow x^*$. Now, using the representation

$$\Psi(\lambda) = M(\lambda) \cap \{x \in R^n \,/\, f(x) \leq \varphi(\lambda)\}$$

the continuity of both the objective function and the constraints $g_i$, i=1,...,m, and the upper semicontinuity of $\varphi$ at $\lambda^0$ one obtains that $x^* \in \Psi(\lambda^0)$ must hold.

The next step will be to prove that $\varphi$ is lower semicontinuous at $\lambda^0$. This, however, needs some preparatory considerations. As in the proof of Theorem 2.4. we define

$$I_1 \underset{Df}{=} \{i \,/\, g_i(x) = \lambda_i \;\forall\, x \in M(\lambda^0)\} ,$$

$$I_2 \underset{Df}{=} \{i \,/\, \exists\, x(i) \in M(\lambda^0) : g_i(x(i)) < \lambda_i^0\}$$

$$M^{I_1}(\lambda^0) \underset{Df}{=} \{x \in R^n \,/\, g_i(x) \leq \lambda_i^0,\; i \in I_1\} ,$$

$$M^{I_2}(\lambda^0) \underset{Df}{=} \{x \in R^n \,/\, g_i(x) < \lambda_i^0,\; i \in I_2\} .$$

we know, by the proof mentioned above, that $I_1 \subset J$ holds and that $M^{I_1}(\lambda^0)$ is an affine subspace coinciding with the space $R^n$ if $I_1 = \emptyset$.
If one now defines

$$\Gamma(u) \underset{Df}{=} u + M^{I_1}(\lambda^0),\; u \in R^n,$$

$$h(u) \underset{Df}{=} \inf \{f(x) \,/\, x \in \Gamma(u) \cap M^{I_2}(\lambda^0) ,$$

then the assumptions of Theorem 1. are fulfilled and, consequently, the value function h is continuous at u = 0. By (2.13) one concludes that $h(0) = \varphi(\lambda^0)$. Hence,

$$\forall \varepsilon > 0 \; \exists x^o \in M^{I_1}(\lambda^o) \cap M^{I_2}(\lambda^o): \varphi(\lambda^o) > f(x^o) - \varepsilon . \qquad (13)$$

Let $\{\lambda^t\} \subset \Lambda$ with $\lambda^t \xrightarrow[I]{} \lambda^o$ and $x^t \in M(\lambda^t)$, for each t be arbitrarily chosen. Since $M^{I_1}(\lambda^o)$ is an affine subspace the corresponding multifunction $M^{I_1}: \Lambda \longrightarrow 2^{R^n}$ is continuous at $\lambda^o$ according to Theorem 2.3. Therefore, there is a sequence $\{u^t\} \subset R^n$ with $x^t \in \Gamma(u^t)$ for all t and $\|u^t\| \longrightarrow 0$. The points

$$z^t \underset{Df}{=} \frac{1}{2}(x^t + x^o)$$

belong to the set $\Gamma(\frac{1}{2} u^t)$. Because of

$$g_i(z^t) \leq 1/2 g_i(x^o) + 1/2 g_i(x^t) < \frac{1}{2}(\lambda_i^o + \lambda_i^t)$$

and

$$g_i(x^o) < \lambda_i^o, \quad i \in I_2, \text{and } g_i(x^t) \leq \lambda_i^t, \; i=1,\dots,m,$$

we have, for large numbers t, that

$$g_i(x^t) < \lambda_i^o.$$

Hence, $z^t \in \Gamma(\frac{1}{2}u^t) + M^{I_2}(\lambda^o)$. Thus, the continuity of h at 0 and $\|u^t\| \longrightarrow 0$ imply that

$$f(z^t) > h(0) - \varepsilon = \varphi(\lambda^o) - \varepsilon$$

holds for large t. By (13) and the convexity of f one gets

$$\begin{aligned} f(x^t) &\geq 2 f(z^t) - f(x^o) > 2(\varphi(\lambda^o) - \varepsilon) - f(x^o) = \\ &= \varphi(\lambda^o) + (\varphi(\lambda^o) - f(x^o)) - 2\varepsilon > \\ &> \varphi(\lambda^o) - 3\varepsilon . \end{aligned}$$

This shows the lower semicontinuity of $\varphi$ at $\lambda^o$, and the proof of (i) is complete.

The proposition (ii) one obtains by Theorem 2.4. if one represents the optimal sets $\Psi(\lambda)$ by

$$\Psi(\lambda) = \{x \in R^n \; / \; f(x) \leq \varphi(\lambda), \; g_i(x) \leq \lambda_i, \; i=1,\dots,m\}$$

and takes into account that $\varphi$ is continuous at $\lambda^o$. #

Theorem 5. shows that if $P(\lambda)$ is of the form above and convexity is required for all the functions, then one can omit interior point conditions if additionally all functions occurring in $P(\lambda)$ are weakly analytic.

Now we cite some recent results for the parametric problem

$$P(p,\lambda) \quad \inf\{f(x) + p^T x \; / \; x \in M(\lambda)\} \; , \; (p,\lambda) \in R^n \times R^m,$$

where $f : R^n \longrightarrow R$ is a fixed polynomial function and $M(\lambda) \underset{Df}{=} \{x \in R^n \; / \; Ax \leq \lambda\}$ with a fixed (m,n)-matrix A. Let the following notations be introduced:

$$\varphi(p,\lambda) \underset{Df}{=} \inf_{x \in M(\lambda)} f(x) + p^T x, \qquad \mathcal{E} \underset{Df}{=} \{(p,\lambda) \in R^n \times R^m \; / \; \varphi(p,\lambda) > -\infty\}$$

$$E(\lambda) \underset{Df}{=} \{p \in R^n \,/\, (p,\lambda) \in \mathcal{E}\}, \quad \Lambda \underset{Df}{=} \{\lambda \in R^m \,/\, E(\lambda) \neq \emptyset, M(\lambda) \neq \emptyset\},$$

$$\Lambda^* \underset{Df}{=} \{\lambda \in \Lambda \,/\, \exists\, \lambda^* \in \Lambda : \lambda < \lambda^*\}.$$

ANDRONOW/BELOUSOV (1985) have shown that $\varphi/_{E(\lambda)}$ is continuous and $\Psi/_{E(\lambda)}$ is u.s.c. (i.e. the case $\lambda$= const. in $P(p,\lambda)$) if f is of a degree $\leq 3$. BELOUSOV (1986) showed that for polynomials f of degree $\leq 3$ it holds:

(i) $\varphi/_{\bar{\mathcal{E}}}$ is continuous and $\Psi/_{\bar{\mathcal{E}}}$ is u.s.c. where $\bar{\mathcal{E}} \underset{Df}{=} \{(p,\lambda) \,/\, \lambda \in \Lambda^*\}$;

(ii) if $S \subset \mathcal{E}$ is a bounded set such that $\exists\, \lambda^0 \in \Lambda^* : (p,\lambda) \in S \Rightarrow \lambda \leq \lambda^0$ then $\varphi/_S$ is Lipschitzian.

Further he could prove that in cases where f is of a degree $\leq 2$ or convex or concave, $\varphi$ is Lipschitzian on a bounded set $S \subset \mathcal{E}$.
For a more general discussion of local stability properties the reader is referred to KLATTE (1984).

The next two theorems deal with a parametric problem of the form

$$P(\lambda) \quad \inf \{f(x,\lambda) \,/\, x \in M\}, \qquad \lambda \in R^m, \tag{14}$$

where the constraint set $M \subset R^n$ is non-empty and fixed and $f : M \times R^m \longrightarrow R$ is an affine linear function for each fixed $x \in M$. The value function of $P(\lambda)$ is again denoted by $\varphi$ and the optimal set-function by $\Psi$. Further we use the definition

$$E \underset{Df}{=} \{\lambda \in R^m \,/\, \varphi(\lambda) > -\infty\}.$$

The following results hold under fairly weak assumptions on the constraint set M (i.e. at most closedness). Thus, they may directly be applied to mixed-integer problems with fixed constraint sets.

Theorem 6.

(i) E is a convex set and the value function $\varphi$ is concave on E and, consequently, continuous on ri E.

(ii) If f is upper semicontinuous on M x E, then $\varphi$ is upper semicontinuous on E.

(iii) If M is a closed set and f is continuous on M x E, then $\Psi$ is closed on E.

Proof.

(i) is clear by the assumption that f(x,.) is an affine linear function for every $x \in M$ and by the well-known continuity property of a concave function.

(ii) Let $\lambda^0 \in E$ be arbitrarily chosen and let $\{\lambda^t\} \subset E$ be a sequence with $\lambda^t \longrightarrow \lambda^0$. For an arbitrary $\varepsilon > 0$ and a point $x^0 \in M$ such that

$f(x^o,\lambda^o) < \varphi(\lambda^o) + \varepsilon$ and an arbitrary sequence $\{x^t\} \subset M$ with $x^t \longrightarrow x^o$ one has

$$\overline{\lim}\, \varphi(\lambda^t) \le \overline{\lim}\, f(x^t,\lambda^t) \le f(x^o,\lambda^o) < \varphi(\lambda^o) + \varepsilon .$$

(iii) Let $\lambda^o \in E$, $\{\lambda^t\} \subset E$ and $\{x^t\} \subset M$ be arbitrarily chosen such that

$$\lambda^t \longrightarrow \lambda^o \text{ and } x^t \in \Psi(\lambda^t) \text{ with } x^t \longrightarrow x^o.$$

Then one has $x^o \in M$ and $f(x^o,\lambda^o) = \overline{\lim}\, f(x^t,\lambda^t) = \overline{\lim}\, \varphi(\lambda^t) \le \varphi(\lambda^o)$, from which it follows that $x^o \in \Psi(\lambda^o)$. #

If the constraint set M in Theorem 6.(iii), in addition, is bounded, then the u.s.c.-property of $\Psi$ is an immediate consequence. Because of the continuity of $\varphi$ and

$$\Psi(\lambda) \subset N(\lambda,\alpha) \underset{Df}{=} \{x \in M \,/\, f(x,\lambda) \le \alpha\}$$

where $\alpha \in R$ with $N(\lambda,\alpha) \ne \emptyset$ it is clear that $\Psi$ is u.s.c. if the *level-set-mapping* N defined by $N(\lambda,\alpha)$ is u.s.c,. In the following we treat a special form

$$f(x,\lambda) = g(x) + \lambda^T x \tag{15}$$

of the objective function for which $N(\lambda,\alpha)$ are compact sets.
Let

$$N(\lambda,\alpha) = \{x \in M \,/\, g(x) + p^T x \le \alpha\},$$

$$\mathcal{E}^o \underset{Df}{=} \{(\lambda,\alpha) \in R^{n+1} \,/\, \lambda \in \text{int } E,\ N(\lambda,\alpha) \ne \emptyset\}.$$

Theorem 7.

Let M be a closed set and g a lower semicontinuous function on M and, let int E be non-empty. If $(\lambda^t, a_t) \in \mathcal{E}^o$, t=1,2,..., is a sequence with $(\lambda^t, a_t) \longrightarrow (\lambda^o, a_o)$ where $\lambda^o \in \text{int } E$, then we have for the level sets of the problem (14), (15):

$$\forall \varepsilon > 0\ \exists t(\varepsilon): N(\lambda^t, a_t) \subset U_\varepsilon N(\lambda^o, a_o)\ \forall t > t(\varepsilon).$$

Proof. Let $\{x^t\} \subset M$ be an arbitrary sequence with $x^t \in N(\lambda^t, a_t)$. If this sequence is convergent (without loss of generality *) to an $x^o \in M$, then one has

$$\alpha_o = \lim \alpha_t \ge \underline{\lim}\, (g(x^t) + \lambda^{tT} x^t) \ge g(x^o) + \lambda^{oT} x^o,$$

from which it follows that $x^o \in N(\lambda^o, a_o)$.
Let us suppose without loss of generality * that

$$\|x^t\| \longrightarrow \infty \text{ and } \frac{x^t}{\|x^t\|} \longrightarrow x^*.$$

We may suppose that $\{\lambda^t\} \subset U_\delta(\lambda^o)$ for a $\delta > 0$ such that still $U_{3\delta}(\lambda^o) \subset \text{int } E$.
Now, for the sequence

$$\bar{\lambda}^t \underset{Df}{=} \lambda^t - \delta \frac{x^t}{\|x^t\|}$$

one obtains that

$$\varphi(\bar{\lambda}^t) \leq g(x^t) + \bar{\lambda}^t x^t = g(x^t) + \lambda^{tT} x^t - \delta \|x^t\| \leq \alpha_t - \delta \|x^t\|$$

holds, from which $\varphi(\bar{\lambda}^t) \longrightarrow -\infty$ follows.

But, this contradicts the continuity of $\varphi$ on int E and the choice of $\lambda^0$ because of

$$\lambda^t \longrightarrow \lambda^0 - \delta x^* \in U_{3\delta}(\lambda^0) \subset \text{int } E. \qquad \#$$

Corollary 1. If the assumptions of Theorem 7. are fulfilled, then the level set $N(\lambda^0, a_0)$ is non-empty.

Corollary 2. If M is a closed set, g a lower semicontinuous function on M and int E $\neq \emptyset$, then $\Psi(\lambda) \neq \emptyset$ for all $\lambda \in$ int E and the optimal set-function is u.s.c. on int E.

## 4. Quasiconvex Polynomial Optimization Problems

In this chapter we focus our interest on constraint sets

$$M(\lambda) = \{x \in R^n \,/\, p_i(x) \leq \lambda_i,\ i = 1,\dots,m\},\ \lambda \in R^m,$$

given by a finite system of quasiconvex polynomial functions $p_i$ and variable right-hand sides. In the first section such properties of quasiconvex polynomial functions, which we shall use in the analysis of the corresponding constraint sets, are presented. Thereby we follow a concept similar to that due to BELOUSOV (1977) for the case of convex polynomial functions. The second part is dedicated to a characterizatior of the sets $M(\lambda)$ and the corresponding multifunction M. In particular we deal with the points:

- Continuity properties of M,
- Representability of $M(\lambda)$ as a sum of a compact set $C(\lambda)$ and a fixed polyhedral cone,
- Identification of such inequalities $p_i(x) \leq \lambda_i$ which are essential for the existence of (mixed-) integer points in $M(\lambda)$.

As an extension of Chapter 3. we summarize in the third section stability results for the particular case of parametric problems with a fixed quasiconvex polynomial objective function and constraint sets $M(\lambda)$ as considered in the second part.

### 4.1. Properties of Quasiconvex Polynomial Functions

Lemma 1.
Let $p(x)$, $x \in R^n$, be a quasiconvex polynomial function. Then the following holds:

(i) If $p(x)$ has the representation

$$p(x) = q_m(x) + q_{m-1}(x) + \dots + q_o$$

where the functions $q_i(x)$, $i = 1,\dots,m$, are forms of the order i, then the form $q_m(x)$ is quasiconvex.

(ii) Let $x^o$, $x^1 \in R^n$, be two fixed points and $u \in R^n$, $u \neq 0$. Then the polynomial functions

$$p_o(t) \underset{Df}{=} p(x^o+tu) \text{ and } p_1(t) = p(x^1+tu),$$

defined on R, are quasiconvex polynomial functions having the same degree d. If $d \geq 1$, then, in both polynomial functions, the same coefficient appears at the power $t^d$.

Proof. (i) Let $x, y \in R^n$ and $\alpha \in (0,1)$ be arbitrarily fixed. Then

$$p(t(\alpha x+(1-\alpha)y)) = p(\alpha(tx) + (1-\alpha)(ty)) \leq$$
$$\leq \max\{p(tx),\ p(ty)\}$$

hold for an arbitrary number $t \in R$. Thus, if $t > 0$, then

$$\frac{p(t(\alpha x+(1-\alpha)y))}{t^m} \leq \max \left\{ \frac{p(tx)}{t^m} + \frac{p(ty)}{t^m} \right\} .$$

According to the representation of p we obtain

$$q_m(\alpha x+(1-\alpha)y) \leq \max \{q_m(x), q_m(y)\}$$

if one takes $t \longrightarrow \infty$.

(ii) Obviously, we have to show the proposition for the case that $x^o \notin \{x/x = x^1 + tu, t \in R\}$. Assume the representations

$$p_o(t) = a_{m_o} t^{m_o} + \ldots + a_o, \quad a_{m_o} \neq 0,$$

$$p_1(t) = b_{m_1} t^{m_1} + \ldots + b_o, \quad b_{m_1} \neq 0.$$

It is clear that these polynomial functions are quasiconvex on R. Let $\alpha \in (0,1)$ be arbitrarily fixed and points $c(\alpha)$ and $d(\alpha)$ be chosen such that

$$x^1 = \alpha x^o + (1-\alpha)\, c(\alpha),$$
$$x^o = \alpha x^1 + (1-\alpha)\, d(\alpha)$$

hold. Then the quasiconvexity of p permits , for each real number t, the estimates

$$p(x^1+\alpha tu) \leq \max \{p(x^o+tu), p(c(\alpha))\}$$

and

$$p(x^o+\alpha tu) \leq \max \{p(x^1+tu), p(d(\alpha)\} ,$$

which can also be written in the form

$$b_{m_1} \alpha^{m_1} t^{m_1} + \ldots \leq \max \{a_{m_o} t^{m_o} + \ldots; p(c(\alpha))\} \tag{1}$$

and

$$a_{m_o} \alpha^{m_o} t^{m_o} + \ldots \leq \max \{b_{m_1} t^{m_1} + \ldots; p(d(\alpha))\}. \tag{2}$$

Since these inequalities must also be valid for the case $|t| \longrightarrow \infty$, one obtains from (1) and (2) that $d = m_o = m_1$ must hold.

Let now $d \geq 1$. Then one concludes from (1) and (2) that $\operatorname{sign} a_d = \operatorname{sign} b_d$ must be true. If d is an even number, then, by the quasiconvexity of p, it follows that $a_d > 0$ and $b_d > 0$.
We still have to show that $a_d = b_d$. We discuss one case only; the others can be considered analogously. Assume that $0 < b_d < a_d$. Now let $\alpha \in (0,1)$ be chosen such that still $a_d \alpha^d > b_d$ holds, which leads for large t to a contradiction to (2), i.e. this case is impossible. #

Remark 1.

If a quasiconvex polynomial function p(x) is constant (or strictly decreasing or strictly increasing) on a line

$S_o = \{x \in R^n \ / \ x = x^o + tu, \ t \in R\}$ (i.e. p has the respective property as a polynomial function of $t \in R$), then p also shows this property on each line $S = \{x \in R^n \ / \ x = x^1 + tu, \ t \in R\}$ which is a parallel line to $S_o$

Lemma 2.

Let $q_m(x)$, $x \in R^n$, be a quasiconvex form of the order $m \geq 1$. Further let us define

$$L \underset{Df}{=} \{x \in R^n \ / \ q_m(x) = 0\}$$

and

$$H \underset{Df}{=} \{x \in R^n \ / \ q_m(x) \leq 0\} .$$

Then the following statements hold:

(i) L is a linear subspace of $R^n$.

(ii) L = H if the order m is an even number.

(iii) If the order m is an odd number, then L is a hyperplane and H is a half-space, both in $R^n$.

Proof. (i) Let U be the linear subspace of maximal dimension satisfying $U \subset L$. We shall show that U = L holds. We assume that there is an element $x^o \in L$ not belonging to U. Obviously, also $-x^o \in L$. Because of the quasiconvexity of q and the property $q_m(0) = 0$ one has that $q_m(tx^o) = 0$ for all $t \in R$. According to Remark 1. the polynomial function $q_m(x)$ is constant on every line S which is parallel to the line $G_o = \{x \in R^n \ / \ x = tx^o, \ t \in R\}$. This means that the linear hull of $U \cup \{x^o\}$ also belongs to L, contradicting the choice of U.

(ii) If there were a point $x \in R^n$ fulfilling $q_m(x) < 0$, then $q_m(-x) < 0$ would hold, too. However, since $q_m$ is quasiconvex, we have

$$0 = q_m(0) = q_m(\tfrac{1}{2}x - \tfrac{1}{2}x) \leq \max \{q_m(x), \ q_m(-x)\} .$$

(iii) We show that the orthogonal complement $L^{\perp}$ of L has the dimension 1. To see this it suffices to assume that $L^{\perp}$ contains two linearly independent vectors $x^1$ and $x^2$. Without loss of generality, we suppose $q_m(x^1) < 0$ and $q_m(x^2) > 0$ (otherwise we would choose $-x^1$ instead of $x^1$, for example). Then there exists an $\alpha \in (0,1)$ such that $q_m(\alpha x^1 + (1-\alpha)x^2) = 0$. The linear independence of $x^1$ and $x^2$ gives $\alpha x^1 + (1-\alpha)x^2 \neq 0$, which leads to a contradiction to

$$L \cap L^{\perp} = \{0\} .$$

Finally, let $x^* \in L^{\perp}$ with $q_m(x^*) < 0$, then by Remark 1. it holds

$$H = \{x \in R^n \ / \ x^{*T}x \geq 0\} . \quad \#$$

Lemma 3.

If the quasiconvex form $q_m(x)$, $x \in R^n$, has only integer coefficients, then the linear subspace

$$L = \{x \in R^n \;/\; q_m(x) = 0\}$$

may be described by a system of homogeneous linear equations with integer coefficients.

Proof. The proof is by induction on the number n of variables appearing in the quasiconvex form $q_m(x)$.

Obviously, the cases n = 0 and n = 1 are trivial. Therefore, let $n \geq 2$ and we suppose that $q_m(x)$ depends on $x_1$.

For a $u \in L$, $u \neq 0$, we have

$$q_m(tu) = t^m q_m(u) = 0 \quad \forall t \in R,$$

and by Lemma 1.

$$q_m(x+u) = q_m(x) \quad \forall x \in R^n \tag{3}$$

holds. The form $q_m$ may be written in a lexicographical manner, i.e.

$$q_m(x) = a_{11}x_1^m + x_1^{m-1}\left(\sum_{j=2}^{n} a_{1j}x_j\right) + $$
$$+ x_1^{m-2}( \ldots ) + \ldots + a_{nn}x_n^m.$$

The coefficient $a_{11}$ must be different from zero, because of Lemma 1. and our assumption that $q_m$ depends on $x_1$. Substituting now x+u for x one obtains

$$q_m(x+u) = q_m(x) + x_1^{m-1}(ma_{11}u_1 + \sum_{j=2}^{n} a_{ij}u_j) +$$
$$+ \ldots + q_m(u).$$

Taking (3) into account, one concludes that

$$x_1^{m-1}(ma_{11}u_1 + \sum_{j=2}^{n} a_{ij}u_j) + \ldots + q_m(u) \equiv 0.$$

This implies, in particular, that

$$ma_{11}u_1 + \sum_{j=2}^{n} a_{1j}u_j = 0, \quad u \in L, \tag{4}$$

must hold. (4) represents a linear subspace U which contains L and which has an integer basis because all coefficients appearing in (4) are integer numbers. This basis can be extended to an integer basis of the space $R^n$. Now, the original quasiconvex form $q_m(x_1,\ldots,x_n)$ can be represented by a quasiconvex form $p_m(y_1,\ldots,y_{n-1})$ with integer coefficients on U. Proceeding now by induction on n we obtain the proposition. #

Remark 2.

Under the hypotheses of Lemma 3., the half-space $H = \{x \in R^n \ / \ q_m(x) \leq 0\}$ (cf. Lemma 2. (iii)) can also be described by a linear inequality with integer coefficients. Such an inequality one constructively finds as sketched above.

Lemma 4.

Let $p(x) = q_m(x) + \dots + q_1(x) + q_0$, $x \in R^n$, be a quasiconvex polynomial which has the degree $\deg(p) = m \geq 1$. Further, let the sets

$$K(p) \underset{Df}{=} \{u \in R^n \ / \ \sup\{p(tu) \ / \ t \geq 0\} < \infty\},$$

$$R(p) \underset{Df}{=} \{u \in R^n \ / \ \sup\{p(tu) \ / \ t \in R\} < \infty\},$$

and an index

$$i_0 \underset{Df}{=} \begin{cases} \max\{i \ / \ \exists \, u \ R^n: \ q_m(u) = \dots = q_{i+1}(u) = 0, \\ \qquad\qquad q_i(u) < 0\}, \text{ if it exists} \\ 1, \text{ otherwise.} \end{cases}$$

be defined.

Then we have

$$K(p) = \{u \in R^n \ / \ q_m(u) = \dots = q_{i_0+1}(u) = 0, \ q_{i_0}(u) \leq 0\},$$

$$R(p) = \{u \in R^n \ / \ q_m(u) = \dots = q_{i_0}(u) = 0\}.$$

Proof. Let $u^0 \in K(p)$. The polynomial function $p_0(t) \underset{Df}{=} p(tu^0)$ on R has the form $p_0(t) = q_d(u^0)t^d + \dots + q_1(u^0)t + q_0$ where $d = \deg(p_0)$. If $d \geq 1$, then $q_d(u^0) < 0$ must hold (if p is a convex polynomial function, then $d = 1$ follows). Therefore,

$$u^0 \in \{u \in R^n \ / \ q_m(u) = \dots = q_{i_0+1}(u) = 0, \ q_{i_0}(u) \leq 0\}.$$

Let now $u^0$ be chosen in the latter set.

Obviously, if $i_0 = 1$ or $q_{i_0}(u^0) < 0$, then $u^0 \in K(p)$ holds. Therefore, let us assume, that $q_{i_0}(u^0) = 0$ and $i_0 > 1$ hold. Because of the maximality of the index $i_0$ the set

$$Q_{i_0} \underset{Df}{=} \{u \in R^n \ / \ q_m(u) = \dots = q_{i_0+1}(u) = 0, \ q_{i_0}(u) < 0\}$$

cannot be empty. Let $u^1 \in Q_{i_0}$.

According to Remark 1., it suffices to show that $p_0(t) = p(tu^0)$ is not increasing for $t \geq 0$, i.e.

$$p(tu^0) \leq p(0) \quad \forall t \geq 0.$$

To see this, we assume that there is a $t_0 > 0$ such that $p(t_0u^0) > p(0)$. According to the definitions of the points $u^0$ and $u^1$ one has, by Lemma 2, that

$$u(\alpha) \underset{Df}{=} \alpha u^1 + (1-\alpha)(t_o u^o) \in Q_{i_o} \quad \forall \alpha \in (0,1]$$

must hold. For every fixed $\alpha$, the polynomial function

$$p_1(t) \underset{Df}{=} p(tu(\alpha)), \quad t \in R,$$

which decreases because of $q_{i_o}(u(\alpha)) < 0$, is quasiconvex. This means that $p(u(\alpha)) < p(0)$ for all $\alpha^o \in (0,1]$. Letting $\alpha \longrightarrow 0$ we obtain by the continuity of p that $p(t_o u^o) \leq p(0)$ must be true, which contradicts the above assumption.

The remaining statement on the set R(p) one sees easily since the equivalence

$$u \in R(p) \Longleftrightarrow u, -u \in K(p)$$

holds. #

Remark 3.

If the polynomial function p(x) is convex, then one obtains BELOUSOV's result $i_o \equiv 1$ in Lemma 4.

The non-zero vectors u of K(p) and R(p) are the directions of all half-lines of the space $R^n$ on which the quasiconvex polynomial function p(x) decreases or is constant, respectively. The following Lemma 5. will give a more precise characterization of the set K(p) and, in particular, it will show that all level sets of a quasiconvex polynomial function have the same recession cone, which is not true for general quasiconvex functions.

Lemma 5.

For a quasiconvex polynomial function p(x), $x \in R^n$, the following statements hold:

(i) K(p) is the recession cone of all non-empty level sets

$$N_\alpha(p) \underset{Df}{=} \{x \in R^n \;/\; p(x) \leq \alpha\}$$

of p(x).

(ii) K(p) is a polyhedral cone.

(iii) If all the coefficients of the polynomial function p(x) are rational numbers, then there is a finite system of linear homogeneous inequalities with only integer coefficients whose solution set is the cone K(p).

Proof. By Lemma 4. the statement (i) follows immediately from the definition of the recession cone $0^+N_\alpha(p)$.

(ii). Using the representation

$p(x) = q_m(x) + \ldots + q_1(x) + q_o$, deg(p) m $\geq$ 1 for the polynomial function p(x) and the successive application of Lemma 1.(i) and

Lemma 2. one obtains the proposition as follows. The set $\{u \in R^n \ / \ q_m(u) = 0\}$ is a linear subspace of $R^n$, the set $\{u \in R^n \ / \ q_m(u) = q_{m-1}(u) = 0\}$ is a linear subspace of $\{u \in R^n \ / \ q_m(u)=0\}$ and so on. Finally, the set K(p) is a half-space of $\{u \in R^n \ / \ q_m(u) = \ldots = q_{i_0}(u) = 0\}$ or coincides with this set, respectively.

(iii). This proposition one obtains by the same procedure as used under (ii) if one additionally takes Remark 2. into account. #

Remark 4.

The proof of Lemma 3. sketches, with respect to Remark 2., an idea how one can derive an algorithm which generates a linear system of inequalities as desired under (iii) of Lemma 5. for a quasiconvex polynomial function with only integer coefficients.

## 4.2. Multifunctions and Constraint Sets Defined by Quasiconvex Polynomial Functions

In this section we shall consider constraint sets of the form (2.10) where all the involved functions are quasiconvex polynomial functions, i.e.

$$M(\lambda) = \{x \in R^n \ / \ p_i(x) \leq \lambda_i, \ i = 1,\ldots,m\} \ ; \ \lambda \in R^m, \tag{5}$$

$p_i : R^n \longrightarrow R$, $i = 1,\ldots,m$, are quasiconvex polynomial functions.

The related constraint set-mapping M is defined on

$$\Lambda \underset{Df}{=} \{\lambda \in R^m \ / \ M(\lambda) \neq \emptyset\}.$$

The next lemma is concerned with the properties of the recession cone of $M(\lambda)$ given by (5).

Lemma 6.

Let $M(\lambda)$ be given by (5), then

$$V \underset{Df}{=} \bigcap_{i=1}^{m} K(p_i) = 0^+M(\lambda) \quad \forall \lambda \in \Lambda. \tag{6}$$

V is a polyhedral cone which is integerly generated if the coefficients of all quasiconvex polynomial functions $p_i$, $i=1,\ldots,m$, are rational numbers.

Proof. Obviously, for all $\lambda \in \Lambda$ it holds

$$0^+M(\lambda) = \bigcap_{i=1}^{m} 0^+\{x \in R^n \ / \ p_i(x) \leq \lambda_i\} = \bigcap_{i=1}^{m} K(p_i) = V.$$

The rest follows immediately by the lemmas of section 4.1. #

Theorem 1.

The constraint set-mapping M defined by (5) is l.s.c.

Proof. Let $\lambda^o \in \Lambda$ be an arbitrary point. We define

$$I \underset{Df}{=} \{i \;/\; p_i(x) = \lambda_i^o,\ \forall x \in M(\lambda^o)\},$$

$$\Gamma(\lambda) \underset{Df}{=} \{x \in R^n \;/\; p_i(x) \leq \lambda_i,\ i \in I\}$$

and

$$M^o(\lambda) \underset{Df}{=} \{x \in R^n \;/\; p_i(x) < \lambda_i,\ i \notin I\}.$$

Then, by Theorem 2.2. it suffices to show that the multifunction $\Gamma$ defined by $\Gamma(\lambda)$ is l.s.c. at $\lambda^o$. Using Remark 1. one easily verifies that the choice of I causes that $\Gamma(\lambda^o)$ is an affine subspace. From Lemma 6. we conclude

$$\Gamma(\lambda) = (\Gamma(\lambda) \cap \bar{V}^{\perp}) + \bar{V}$$

where $\bar{V}$ is the linear subspace defined by

$$\bar{V} \underset{Df}{=} 0^+\Gamma(\lambda^o).$$

Obviously, it suffices now to show that the multifunction $\bar{\Gamma}$, given by

$$\bar{\Gamma}(\lambda) \underset{Df}{=} \Gamma(\lambda) \cap \bar{V}^{\perp}$$

is l.s.c. at $\lambda^o$. Since the sets $\bar{\Gamma}(\lambda)$ are compact, for all $\lambda \in \Lambda$, the multifunction $\bar{\Gamma}$ is u.s.c. at $\lambda^o$, by Theorem 2.1. Taking into account that $\bar{\Gamma}(\lambda^o)$ is a single point, it is clear that $\bar{\Gamma}$ is l.s.c. at $\lambda^o$, too. #

Remark 5.

The constraint set-mapping M is continuous if V is a linear subspace, which becomes clear by the above proof.

The following counterexample (due to SHIRONIN) shows that the mapping M is not u.s.c. in general not even in the case of convex polynomial functions. For an analysis of this example see BANK/BELOUSOV et al. (1986). The mapping M defined by

$$M(\lambda) \underset{Df}{=} \{x \in R^4 \;/\; p_1(x) \leq \lambda_1,\ p_2(x) \leq \lambda_2,\ x_2^2 \leq \lambda_3,\ -x_1 \leq \lambda_4,\ -x_3 \leq \lambda_5\}$$

with

$$p_1(x) = x_1^{16} + 2x_2^8 + x_2^7 + 2x_2^2 + 2x_3^6 + x_3^5 + 2x_3^2 - x_4 + x_1x_2^3x_3^3 + x_1^2x_2^4x_3^2 + x_2^2x_3^4 + x_1^4x_3^4 + x_1^4x_2^6 + x_1^2x_2^6,$$

$$p_2(x) = 2x_2^8 + 2x_3^6 + x_2^2x_3^4 + 2x_2^2 + 2x_3^2 - x_4$$

on

$$\Lambda = \{\lambda \in R^5 \;/\; M(\lambda) \neq \emptyset\} = \{\lambda \in R^5 \;/\; \lambda_3 \geq 0\}$$

is not u.s.c. at $\lambda$ if $\lambda_3 = 0$. The reason for that consists in the fact that the polynomial function $p_1(x)$ is not a so-called primitive one (see Theorem 6.).

Lemma 7.

Let V be the recession cone (6) and let $u \in V$, $u \neq 0$, be fixed. Further, let

$$H \underset{Df}{=} \{x \in R^n \;/\; u^T x = 0\} .$$

Then the projection $P_H(M(\lambda))$ of the constraint set $M(\lambda) \neq \emptyset$ on the hyperplane H is given by

$$P_H(M(\lambda)) = \begin{cases} H, \text{ if } I = \emptyset \\ \{x \in H \;/\; p_i(x) \leq \lambda_i,\; i \in I\}, \text{ otherwise} \end{cases} \tag{7}$$

where $I \underset{Df}{=} \{i \;/\; u \in R(p_i)\}$.

Proof. We consider the case $I \neq \emptyset$. If $x \in M(\lambda)$, then there is a $t \in R$ such that $x + tu \in H$. The choice of I causes that

$$p_i(x+tu) = p_i(x) \leq \lambda_i, \quad \forall i \in I \tag{8'}$$

holds.

Let now $x \in H$ with $p_i(x) \leq \lambda_i \;\; \forall i \in I$. Then (8') holds for all $t \in R$. Since u belongs to the recession cone $0^+M(\lambda)$, every polynomial function $q_i(t) \underset{Df}{=} p_i(x+tu)$, $i \in I$, is decreasing for $t \geq 0$. Hence, for sufficiently large t, it follows that $x + tu \in M(\lambda)$. #

For further characterizations of the constraint set-mapping M given by (5) we need a certain knowledge about the infimum $\varphi(\lambda)$ of a quasi-convex polynomial function p(x) on $M(\lambda)$ for all $\lambda \in \Lambda$, i.e.

$$\varphi(\lambda) \underset{Df}{=} \inf\{p(x) \;/\; x \in M(\lambda)\} , \quad \lambda \in \Lambda . \tag{8}$$

Lemma 8.

If u is a non-zero direction of recession of $M(\lambda)$ (5) and the quasi-convex polynomial function p is constant on a half-line generated by u, i.e.

$$u \in 0^+M(\lambda) \cap R(p),$$

then it holds for the infimum (8) that

$$\varphi(\lambda) = \inf\{p(x) \;/\; x \in P_H(M(\lambda))\} , \lambda \in \Lambda,$$

where $P_H(M(\lambda))$ is defined by (7).

Proof. The proposition follows immediately from Remark 1. taking Lemma 7. into account. #

Theorem 2.

If $p(x)$, $x \in R^n$, is a quasiconvex polynomial function and the constraint set $M(\lambda)$ given by (5), then the value function $\varphi(\lambda)$ (8) is either continuous on $\Lambda$ or $-\infty$ for all $\lambda \in \Lambda$.

Proof.

First we show by contradiction that

$$\exists \bar{\lambda} \in \Lambda: \ \varphi(\bar{\lambda}) = -\infty \Longrightarrow \varphi(\lambda) = -\infty \ \forall \lambda \in \Lambda. \qquad (9)$$

holds. We assume that the proposition of (9) is false. Then there exists, among the counterexamples to (9), one with minimal dimension of the recession cone V of the constraint sets and for which $\varphi(\lambda^0) > -\infty$ hold for some $\lambda^0$. Obviously, $\dim V \geq 1$.

Since $\varphi(\bar{\lambda}) = -\infty$, there must be a direction $u \in V$, $u \neq 0$, which belongs to $R(p)$ or to $K(p) \setminus R(p)$.

The first case is impossible because, according to Lemma 8., we could construct another counterexample having a recession cone of smaller dimension than V. In the second case we have

$$S = \{x \in R^n \ / \ x = x^0 + tu, \ t \geq 0\} \subset M(\lambda^0)$$

for a fixed $x^0 \in M(\lambda^0)$ and

$$\inf\{p(x) \ / \ x \in S\} = -\infty$$

contradicts $\varphi(\lambda^0) > -\infty$. Therefore, (9) is true.

The upper semicontinuity of $\varphi$ on $\Lambda$ is an immediate consequence of the continuity of $p(x)$ and the l.s.c. property of M on $\Lambda$ (Theorem 1.).

We have still to show that $\varphi$ is lower semicontinuous on $\Lambda$, which we execute by contradiction. If the proposition is false, then there exists a counterexample for which the dimension of the recession cone V of the sets $M(\lambda)$ is minimal. We consider such an example. Let $\varphi$ be not lower semicontinuous at some $\lambda^0 \in \Lambda$, and let $\{\lambda^t\} \subset \Lambda$ be a sequence such that

$$\lambda^t \longrightarrow \lambda^0 \text{ and } \underline{\lim} \ \varphi(\lambda^t) \leq \varphi(\lambda^0) - \varepsilon \qquad (10)$$

for a certain $\varepsilon > 0$. By Remark 5. $\dim V \geq 1$ must hold. We may assume that $V \cap R(p) = \{0\}$ because otherwise we could construct, using Lemma 8., another counterexample with constraint sets having a recession cone U such that $\dim U < \dim V$. Thus the sets

$$\bar{M}(\lambda) \underset{Df}{=} M(\lambda) \cap \{x \in R^n \ / \ p(x) \leq \varphi(\lambda^0) - \tfrac{\varepsilon}{2}\}, \ \lambda \in \Lambda,$$

are empty or compact. Because of (10), we assume without loss of generality* that $\bar{M}(\lambda^t) \neq \emptyset$. Then the corresponding multifunction $\bar{M}$ is continuous at $\lambda^0$ according to Remark 5.

Therefore, there is a point $x^0 \in \bar{M}(\lambda^0)$ with $p(x^0) \leq \varphi(\lambda^0) - \frac{\varepsilon}{2}$, which contradicts the definition of $\varphi(\lambda^o)$. #

Now we are ready to decide under which conditions the constraint set-mapping M defined by (5) belongs to the class of multifunctions introduced in (2.15).
Our considerations will make use of an implicit representation of the polyhedral recession cone $V = 0^+M(\lambda)$, i.e.

$$V = \{u \in R^n \ / \ d^{j^T} u \leq 0, \ j = 1,\dots,r\}, \tag{11}$$

which is always possible.
It is clear that the conditions

$$\sup\{d^{j^T} x \ / \ x \in M(\lambda)\} < \infty, \ j = 1,\dots,r, \tag{12}$$

are necessary for the representation of the sets $M(\lambda) \neq \emptyset$ as the sum of a compact set and the recession cone V. We shall show that these conditions are also sufficient for such a representation. Above all, we shall see: if the conditions (12) are satisfied for a single point $\lambda^o \in \Lambda$, then there even exists a continuous compact-valued multifunction $C: \Lambda \longrightarrow R^n$ such that

$$M(\lambda) = C(\lambda) + V, \ \forall \lambda \in \Lambda \tag{13}$$

holds.

Theorem 3.
Let $M: \Lambda \longrightarrow 2^{R^n}$ be given by (5). Further, let (12) be satisfied for a $\lambda^o \in \Lambda$. Then there exists a continuous multifunction $C: \Lambda \longrightarrow 2^{R^n}$ such that (13) holds and the sets $C(\lambda), \lambda \in \Lambda$, are compact and convex.

Proof. Let H be the subspace of maximal dimension contained in the recession cone V. Then the closed convex sets $M(\lambda), \lambda \in \Lambda$, can be written in the form

$$M(\lambda) = (M(\lambda) \cap H^{\perp}) + H.$$

If $V \cap H^{\perp} = \{0\}$, then the non-empty sets $M(\lambda) \cap H^{\perp}$ are compact and convex.
By Remark 5., the multifunction C having the latter sets as images is the desired one.
Thus, we have to investigate the case $V \cap H^{\perp} \neq \{0\}$. Our aim is to construct a continuous function $t: \Lambda \longrightarrow R$ such that the multifunction $C: \Lambda \longrightarrow 2^{R^n}$ defined by

$$C(\lambda) \underset{Df}{=} \{x \in M(\lambda) \ / \ a^T x \geq t(\lambda)\} \cap H^{\perp}, \qquad a \underset{Df}{=} \sum_{j=1}^{r} d^j \tag{14}$$

satisfies (13). Then, the choice of the vector a guarantees the compact-

ness of $C(\lambda)$ and the continuity of C follows from the continuity of t and Remark 5.
In order to construct the function t we define

$$\varphi_j(\lambda) \underset{Df}{=} \sup\{d^{j^T}x \ / \ x \in M(\lambda)\}, \ j = 1,\dots,r,$$

$$M^*(\lambda) \underset{Df}{=} \{x \in R^n \ / \ d^{j^T}x \le \varphi_j(\lambda), \ j = 1,\dots,r\}, \lambda \in \Lambda \ .$$

Theorem 2. and the fact that the conditions (12) are satisfied for $\lambda^0$ imply that the functions $\varphi_j$, $j = 1,\dots,r$, are continuous on $\Lambda$. For the polyhedra $M^*(\lambda)$ we have

$$M(\lambda) \subset M^*(\lambda) \text{ and } 0^+M^*(\lambda) = V, \quad \lambda \in \Lambda.$$

As a tool we shall prove that the proposition of our theorem is correct for the multifunction $M^*$ (i.e. the theorem is valid for the case of linear constraints). In order to find a corresponding continuous function $t^*(\lambda)$ we define now

$$z(\lambda) \underset{Df}{=} (\varphi_1(\lambda),\dots,\varphi_r(\lambda))^T \in R^r \ \forall \lambda \in \Lambda,$$

$$D^T \underset{Df}{=} (d^1,\dots,d^r),$$

$$W \underset{Df}{=} \left\{\binom{v}{w} \in R^{2r} \ / \ D^Tv - D^Tw = -a, \ v,w \ge 0\right\},$$

$$U(\lambda,x) \underset{Df}{=} \left\{\binom{\alpha}{u} \in R^{1+n} \ / \ \begin{array}{l} Du \le 0 \\ Du \ge Dx - z(\lambda) \\ a^Tu + \alpha \le a^Tx \end{array}\right\}, \lambda \in \Lambda .$$

For each $\lambda \in \Lambda$ and each $x \in M^*(\lambda)$ we consider the following linear optimization problem:

$$P(\lambda,x) \ \max\left\{\alpha \ / \ \binom{\alpha}{u} \in U(\lambda,x)\right\}.$$

This problem has always an optimal solution since $\binom{a^Tx}{0} \in U(\lambda,x)$ and the objective function is bounded from above on $U(\lambda,x)$. Hence, the dual problem $D(\lambda,x) \ \min \left\{-(Dx-z(\lambda))^Tw + a^Tx \ / \ \binom{v}{w} \in W\right\}$ possesses always an optimal solution, too, and we denote by $h(\lambda,x)$ the common optimal value of both problems.
If $\binom{v^1}{w^1},\dots,\binom{v^q}{w^q}$ are the extremal points of $W$, then (by linear programming)

$$t^*(\lambda) \underset{Df}{=} \min_{i \le i \le q} z(\lambda)^T(w^i - v^i) \le h(\lambda,x). \tag{15}$$

That the minorant $t^*(\lambda)$ of $h(\lambda,x)$ is a continuous function one easily sees from the definition of $z(\lambda)$.
Let

$$C^*(\lambda) \underset{Df}{=} \{x \in M^*(\lambda) \ / \ a^Tx \ge t^*(\lambda)\} \cap H^\perp. \tag{16}$$

Then $C^*(\lambda)$ is compact and convex for each $\lambda$ and it holds

$$M^*(\lambda) = C^*(\lambda) + V \quad \forall\lambda\in\Lambda$$

because of the definition of $P(\lambda,x)$ and (15). This means that $t^*$ is the desired continuous function.

**Now we are proceed with the proof of the general case and we will use the** above notations.

Let $\{u^1,\dots,u^c\}$ be a system of generators of the polyhedral cone $V\cap H^{\perp}$. We use the following definitions:

$$I(p_j) \underset{Df}{=} \{i\in\{1,\dots,c\} / \ u^i \notin R(p_j)\} \ , \ j = 1,\dots,m,$$

$$I \underset{Df}{=} \{j\in\{1,\dots,m\} / \ I(p_j) \neq \emptyset\},$$

$$t_{ij}(\lambda) \underset{Df}{=} \min\{\alpha / \alpha \geqq 0, \ p_j(k+\alpha u^i) \leqq \lambda_j, \ \forall k\in C^*(\lambda)\}$$

for $j\in I$ and $i\in I(p_j)$.

The index set I is non-empty because of the definition of $R(p_j)$ and $V\cap H^{\perp} \neq \{0\}$. We will use the functions $t_{ij}(\lambda)$ to construct the function $t(\lambda)$ in question and, first, we show that each $t_{ij}(\lambda)$ is continuous on $\Lambda$. We assume that $t_{ij}(\lambda)$ is not upper semicontinuous, then we have

$$\exists\{\lambda^l\}\subset\Lambda \ \exists \ \varepsilon>0 : \lambda^l \longrightarrow \lambda^0\in\Lambda \ \text{ and } \ t_{ij}(\lambda^l) > t_{ij}(\lambda^0) + \varepsilon \ , \ l = 1,2,\dots .$$

Hence, there are points $k^l\in C^*(\lambda^0)$ with

$$p_j(k^l+(t_{ij}(\lambda^0) + \tfrac{\varepsilon}{2})u^i) > \lambda_j^l, \ l = 1,2,\dots .$$

Underlying without loss of generality* that $k^l \longrightarrow k^*\in C^*(\lambda^0)$ it follows that

$$p_j(k^* + (t_{ij}(\lambda^0) + \tfrac{\varepsilon}{2})u^i) \geqq \lambda_j^0.$$

Since the definitions of $I(p_j)$ and $i\in I(p_j)$ cause that $p_j(x)$ decreases on $\{x\in R^n / \ x = k^* + \mu u^i, \ \mu \geqq 0\}$ , we have $p_j(k^*+t_{ij}(\lambda^0)u^i) > \lambda_j^0$, which contradicts the definition of $t_{ij}(\lambda^0)$. Therefore, the functions $t_{ij}(\lambda)$ are upper semicontinuous on $\Lambda$.

In order to show that $t_{ij}(\lambda)$ is lower semicontinuous on $\Lambda$ we assume that

$$\exists\{\lambda^l\} \subset \Lambda \ \exists \ \varepsilon>0 : \lambda^l \longrightarrow \lambda^0\in\Lambda \ \text{ and } \ t_{ij}(\lambda^l) < t_{ij}(\lambda^0) - \varepsilon \ , \ l=1,2,\dots,$$

Assume without loss of generality*

$$t_{ij}(\lambda^l) \longrightarrow \bar{t} \leqq t_{ij}(\lambda^0) - \varepsilon .$$

Let $k\in C^*(\lambda^0)$ be arbitrarily fixed. By the l.s.c property of $C^*$ there is a sequence $\{k^l\}: k^l\in C^*(\lambda^l)$ and $k^l \longrightarrow k$. From $p_j(k^l+t_{ij}(\lambda^l)u^i) \leqq \lambda_j^l$ one obtains $p_j(k+\bar{t}u^i) \leqq \lambda_j^0$ and, therefore, it holds

$$p_j(k+\bar{t}u^i) \leqq \lambda_j^0 \quad \forall k\in C^*(\lambda^0).$$

This contradicts the definition of $t_{ij}(\lambda^0)$ because of $\bar{t} \leqq t_{ij}(\lambda^0) - \varepsilon$.
We define the function

$$t_1(\lambda) \underset{Df}{=} \max\{t_{ij}(\lambda) / \ j\in I, \ i\in I(p_j)\}$$

which is continuous on $\Lambda$ and which allows us to describe the desired function t:

$$t(\lambda) \underset{Df}{=} t^*(\lambda) + t_1(\lambda) \sum_{i=1}^{c} a^T u^i, \quad \lambda \in \Lambda.$$

Obviously, $t(\lambda)$ is continuous.

It remains to show now that the sets $C(\lambda)$ defined by (14) have the property (13).

The inclusion $C(\lambda) + V \subset M(\lambda)$, $\lambda \in \Lambda$, is obvious. We verify that the opposite inclusion is valid, too. Let $\lambda \in \Lambda$ and $x \in M(\lambda)$ be arbitrarily fixed. Then the point x can be represented in the form

$$x = k + \sum_{i=1}^{c} \alpha_i u^i + h \quad \text{with } k \in C^*(\lambda),\ h \in H \text{ and } \alpha_1, \dots, \alpha_c \geqq 0.$$

We set

$$\alpha^* = t_1(\lambda),\ k^* = k + \sum_{i=1}^{c} \min(\alpha_i, \alpha^*) u^i \text{ and}$$

$$u^* = \sum_{i=1}^{c} \max(0, \alpha_i - \alpha^*) u^i$$

and obtain the representation

$$x = k^* + u^* + h \text{ with } u^* + h \in V,\ k^* \in H^\perp.$$

Hence it suffices to show

$$a^T k^* \geqq t(\lambda) \text{ and } p_j(k^*) \leqq \lambda_j,\ j = 1, \dots, m.$$

We have

$$a^T k^* = a^T k + \sum_{i=1}^{c} \min(\alpha_i, \alpha^*) a^T u^i \geqq$$

$$\geqq t^*(\lambda) + \sum_{i=1}^{c} (t_1(\lambda) \cdot a^T u^i) = t(\lambda)$$

because of $k \in C^*(\lambda)$, $t_1(\lambda) \geqq \min(\alpha_i, \alpha^*) \geqq 0$ and $a^T u^i \leqq 0$.

For the remaining part of the proof we distinguish between two cases. Let $j \in \{1, \dots, m\}$ be arbitrarily fixed.

<u>Case 1.:</u> $\alpha_i > \alpha^* \Longrightarrow i \notin I(p_j)$.

Here one has

$$p_j(k^*) = p_j(k + \sum_{i \in I(p_j)} \alpha_i u^i + \sum_{i \notin I(p_j)} \min(\alpha_i, \alpha^*) u^i) =$$

$$= p_j(k + \sum_{i \in I(p_j)} \alpha_i u^i + \sum_{i \notin I(p_j)} \alpha_i u^i)$$

$$= p_j(k + \sum_{i=1}^{c} \alpha_i u^i) = p_j(k + \sum_{i=1}^{c} \alpha_i u^i + h) =$$

$$= p_j(x) \leqq \lambda_j.$$

<u>Case 2.:</u> $\exists i^* \in I(p_j) : \alpha_{i^*} > \alpha^*$.

With $\alpha^* \geqq t_{i^*j}(\lambda)$ one obtains

$$p_j(k^*) = p_j(k + \sum_{i=1}^{c} \min(\alpha_i, \alpha^*)u^i) \leq$$
$$\leq p_j(k + \min(\alpha_{i*}, \alpha^*)u^{i*}) =$$
$$= p_j(k + \alpha^* u^{i*}) \leq \lambda_j. \quad \#$$

Remark 6:
In our further considerations we shall work with constraint sets of the form (5) and often we shall need the validity of Theorem 3. This theorem is true if and only if the constraint sets $M(\lambda)$ (5) show the property:

(CPC) There exist a $\lambda^o \in \Lambda$ and a compact set $C_o \subset R^n$ such that $M(\lambda^o)$ is the sum of $C_o$ and the recession cone V of $M(\lambda^o)$.

Corollary 1.
If the constraint sets $M(\lambda)$ defined by (5) satisfy (CPC), then M is H-continuous and the domain $\Lambda$ of M is closed.

Proof. Using the representation (13) the H-continuity of M follows from the compactness of $C(\lambda)$ and the constancy of V. The remaining part is obvious. #

The following example shows that the representation (13) of $M(\lambda)$ for all $\lambda \in \Lambda$ essentially depends on the property of the constraints $p_j(x)$ to be quasiconvex polynomial functions.

Example.
The multifunction M defined by
$$M(\lambda) = \{x \in R^2 \,/\, e^{x_1} \leq x_2,\ x_1 \leq \lambda_1,\ -x_2 \leq \lambda_2\}$$
is continuous on $\Lambda = R^2$. For all $\lambda$ with $\lambda_2 < 0$ a representation of $M(\lambda)$ as a sum of a compact set and the recession cone
$$V = \{u \in R^2 \,/\, u_1 \leq 0,\ u_2 \geq 0\}$$
exists, but, for $\lambda_2 \geq 0$ such a representation is impossible.

Corollary 2.
If the constraint sets $M(\lambda)$ (5) are given by linear functions
$$p_j(x) = d^{j^T} x,\ j = 1,\dots,m, \text{ i.e.}$$
$$M(\lambda) = \{x \in R^n \,/\, d^{j^T} x \leq \lambda_j,\ j = 1,\dots,m\}$$
(obviously, it holds that $0^+M(\lambda) = V = \{u \in R^n \,/\, d^{j^T} u \leq 0,\ j = 1,\dots,m\}$,) then the compact-valued multifunction $C : \Lambda \longrightarrow 2^{R^n}$ according to Theorem 3. can be chosen such that there is a constant number $\rho > 0$ independent of $\lambda$ and it holds

$$\|x\| \leq \varrho \|\lambda\| \quad \forall x \in C(\lambda) \quad \forall \lambda \in \Lambda .$$

Proof.

For the orthogonal complement $H^{\perp}$ used in the proof of Theorem 3. we may underly the representation

$$H^{\perp} = \{x \in R^n \;/\; c^{i^T} x = 0,\ i = 1,\dots,r \leq n\} ,$$

where $\|c^i\| \leq 1$, $i = 1,\dots,r$. The multifunction C may be chosen according to (15) and (16) and we obtain

$$C(\lambda) = \{x \in R^n \;/\; d^{j^T} x \leq \lambda_j,\ j = 1,\dots,m; \quad (\sum_{j=1}^{m} d^j)^T x \geq \min_{1 \leq k \leq q} (w^k - v^k)^T \lambda ; \quad c^{i^T} x = 0,\ i = 1,\dots,r\} ,\ \lambda \in \Lambda . \tag{17}$$

Now, let $\lambda \in \Lambda$ be arbitrarily fixed and the point x a vertex of the polyhedron $C(\lambda)$ (17). Then there are a linearly independent system of vectors $\{a^1,\dots,a^n\}$ contained in the system

$$S \underset{Df}{=} \{d^1,\dots,d^m,\ \sum_{j=1}^{m} d^j,\ c^1,\dots,c^r\}$$

and numbers $\alpha_1,\dots,\alpha_n$ with

$$\alpha_i \in \{\lambda_1,\dots,\lambda_m,\ 0,\ (w^1-v^1)^T\lambda ,\dots,(w^q-v^q)^T\lambda\}$$

such that the vertex x is the only solution of the system of linear equations

$$a^{i^T} x = \alpha_i ,\ i = 1,\dots,n.$$

By CRAMER's rule and HADAMARD's inequality one obtains

$$\|x\|^2 \leq n(n-1) \frac{\|a^*\|^2}{\|a^{**}\|^2} \cdot \|\alpha\|^2 ,$$

where the notations

$$\|a^*\| \underset{Df}{=} \max \{\|b\| \;/\; b \in S\} ,$$

$$\|a^{**}\| \underset{Df}{=} \min \{\|b\| \;/\; b \in S\} ,$$

$$\alpha^T \underset{Df}{=} (\alpha_1,\dots,\alpha_n)$$

are used. Obviously, the numbers $\|a^*\|$ and $\|a^{**}\|$ are independent of $\lambda$ . Using

$$|(w^k-v^k)^T\lambda| \leq \|w^k-v^k\| \cdot \|\lambda\| \quad \forall k \in \{1,\dots,q\}$$

one estimates

$$\|\alpha\| \leq \|\lambda\| \ (1+n \max_{1 \leq k \leq q} \|w^k-v^k\| )$$

and, therefore

$$\|x\| \leq \sqrt{n(n-1)}\ (1+n \cdot \max_{1 \leq k \leq q} \|w^k-v^k\| )\ \frac{\|a^*\|}{\|a^{**}\|} \cdot \|\lambda\| . \tag{18}$$

Putting the coefficient of $\|\lambda\|$ equal to $\rho$ and taking into account that every point $x \in C(\lambda)$ can be represented as a convex-combination of the vertices of the polyhedron $C(\lambda)$ (17) one obtains the proposition. #

Corollary 2. enables us to prove Corollary 3., a known version (WALKUP/WETS (1969), KLATTE (1985)) of a famous lemma of HOFFMAN (1952) where only linear programming methods are applied. Corollary 4. contains a stronger result, which we shall utilize in the considerations for parametric (mixed-) integer linear programming later.

We say that a multifunction $\Gamma : \Lambda \longrightarrow 2^{R^n}$ is Lipschitzian on a subset $\Lambda_0 \subset \Lambda$ if there is a constant number $\rho > 0$ such that

$$d(\Gamma(\lambda), \Gamma(\lambda')) \leq \rho \|\lambda - \lambda'\| \quad \forall \lambda, \lambda' \in \Lambda_0 .$$

Corollary 3.

Under the hypothesis of Corollary 2. the multifunction M is Lipschitzian on $\Lambda$ = dom M.

Proof.

Let $D = (d^{j^T})$, i.e.

$$M(\lambda) = \{x \in R^n \; / \; Dx \leq \lambda\},$$

and let $\lambda, \lambda^* \in \Lambda$ , $x \in M(\lambda)$ be fixed. We define

$$\tilde{M}(\beta) \underset{Df}{=} \{(u,v,y,z) \in R^{2(n+m)} \; / \; Du+v - Dy-z = \beta, \; v,z \geq 0\} \tag{19}$$

having the recession cone

$$\tilde{V} \underset{Df}{=} \{(u,v,y,z) \in R^{2(n+m)} \; / \; Du+v-Dy-z = 0, \; v,z \geq 0\} . \tag{20}$$

Then, by Corollary 2., there is a continuous compact-valued mapping $\tilde{C}$ and a constant number $\eta > 0$ such that

$$\tilde{M}(\beta) = \tilde{C}(\beta) + \tilde{V}, \; \beta \in \text{dom } \tilde{M},$$
$$\|(u,v,y,z)\| \leq \eta \|\beta\| \; \forall (u,v,y,z) \in \tilde{C}(\beta).$$

If $\bar{x} \in M(\lambda^*)$ is a fixed point, then the point

$$(u,v,y,z) = (x, \lambda - Dx, \bar{x}, \lambda^* - D\bar{x})$$

belongs to $\tilde{M}(\lambda - \lambda^*)$. Hence, there are points $(\hat{u},\hat{v},\hat{y},\hat{z}) \in \tilde{C}(\lambda - \lambda^*)$ and $(\tilde{u},\tilde{v},\tilde{y},\tilde{z}) \in \tilde{V}$ such that

$$(x, \lambda - Dx, \bar{x}, \lambda^* - D\bar{x}) = (\hat{u},\hat{v},\hat{y},\hat{z}) + (\tilde{u},\tilde{v},\tilde{y},\tilde{z})$$

and

$$\|(\hat{u},\hat{v},\hat{y},\hat{z})\| \leq \eta \|\lambda - \lambda^*\| .$$

We consider now the point

$$(x^*, v^*) \underset{Df}{=} (\hat{y}, \hat{z}) + (\tilde{u}, \tilde{v})$$

for which we have

$$Dx^* + v^* = D\hat{y} + \hat{z} + D\tilde{u} + \tilde{v}$$

and, by the definition of $\tilde{V}$,

$$Dx^* + v^* = D\hat{y} + \hat{z} + D\tilde{y} + \tilde{z} = D(\hat{y}+\tilde{y}) + (\hat{z}+\tilde{z}) =$$
$$= D\bar{x} + \lambda^* - D\bar{x} = \lambda^*$$

and, since $v^* \geqq 0$, we have $x^* \in M(\lambda^*)$.

Now we estimate

$$\|x-x^*\| = \|(\hat{u}-\tilde{u}) - (\hat{y}+\tilde{u})\| = \|\hat{u}-\hat{y}\| \leqq$$
$$\leqq \|\hat{u}\| + \|\hat{y}\| \leqq 2\eta\|\lambda-\lambda^*\|,$$

and by $\varrho = 2\eta$ the proposition follows. #

Corollary 4.

The multifunction $C : \Lambda \longrightarrow 2^{R^n}$ defined by (17) is Lipschitzian on $\Lambda$.

Proof.

Let $\lambda, \lambda^* \in \Lambda$ be fixed and let

$$\tilde{C}(\lambda, \eta(\lambda)) \underset{Df}{=} C(\lambda) \ , \ x \in C(\lambda),$$

then, by Corollary 3., there are a constant number $\delta > 0$ independent of $\lambda$ and a point $x^* \in \tilde{C}(\lambda^*, \eta(\lambda^*))$ such that

$$\|x-x^*\| \leqq \delta \left\| \begin{pmatrix} \lambda \\ \eta(\lambda) \end{pmatrix} - \begin{pmatrix} \lambda^* \\ \eta(\lambda^*) \end{pmatrix} \right\| .$$

Hence

$$\|x-x^*\| = \delta(\|\lambda-\lambda^*\| + |\eta(\lambda) - \eta(\lambda^*)|) =$$
$$= \delta(\|\lambda-\lambda^*\| + |\max_{1 \leqq k \leqq q} (v^k-w^k)^T\lambda - \max_{1 \leqq k \leqq q} (v^k-w^k)^T\lambda^*|.$$

Now there is a certain $k^* \in \{1,\dots,q\}$ such that

$$|\max_{1 \leqq k \leqq q} (v^k-w^k)^T\lambda - \max_{1 \leqq k \leqq q} (v^k-w^k)^T\lambda^*| \leqq |(v^{k^*}-w^{k^*})^T\lambda - (v^{k^*}-w^{k^*})\lambda^*|$$

holds and one obtains

$$\|x-x^*\| \leqq \delta(1+\|v^{k^*}-w^{k^*}\|)\|\lambda-\lambda^*\|,$$

and setting $\varrho = \delta(1+\max_{1 \leqq k \leqq q} \|v^k-w^k\|)$ the proposition follows. #

Obviously, it cannot be expected that the Lipschitzian property holds for more general cases of the form (5) than the linear one considered above. An alternative way is to look for locally Lipschitzian behaviour of multifunctions.

Let $\Gamma : \Lambda \subset R^m \longrightarrow 2^{R^n}$ be a multifunction. The set

$$\text{graph } \Gamma \underset{Df}{=} \left\{ \begin{pmatrix} \lambda \\ x \end{pmatrix} \in \Lambda \times R^n \ / \ x \in \Gamma(\lambda) \right\}$$

is called the *graph of* $\Gamma$.

A multifunction $\Gamma$ is said to be *locally upper Lipschitzian at* $\lambda^0 \in \Lambda$ if there exist numbers $\delta > 0$ and $\gamma > 0$ such that

$$\Gamma(\lambda) \subset \Gamma(\lambda^0) + \gamma \|\lambda - \lambda^0\| S \quad \forall \lambda \in U_\delta\{\lambda^0\}$$

holds, where S denotes the closed unit ball in $R^n$.

Further, $\Gamma$ is called locally lower Lipschitzian at $\lambda^0 \in \Lambda$ if there exist numbers $\delta > 0$ and $\gamma > 0$ such that

$$d(x, \Gamma(\lambda)) \leq \gamma \|\lambda - \lambda^0\| \quad \forall \lambda \in U_\delta\{\lambda^0\} \forall x \in \Gamma(\lambda^0)$$

holds.

We say that $\Gamma$ is locally Lipschitzian at $\lambda^0 \in \Lambda$ if it is both locally upper and lower Lipschitzian at $\lambda^0$.

$\Gamma$ is said to be locally Lipschitzian on $\Lambda^0 \subset \Lambda$ if it has this property at every point $\lambda \in \Lambda^0$.

KLATTE (1985) has shown the following theorem (in a more general form), which implies locally Lipschitzian properties for the constraint set mapping considered here.

Theorem 4.

If $\Gamma : \Lambda \longrightarrow 2^{R^n}$ ($\Lambda = \operatorname{dom}\Gamma$) is a multifunction with a closed convex graph and $\operatorname{int}\Lambda \neq \emptyset$, then

(i) $\Gamma$ H-l.s.c. at $\lambda^0 \in \operatorname{int}\Lambda \Rightarrow \Gamma$ locally lower Lipschitzian at $\lambda^0$;

(ii) $\Gamma$ H-continuous at $\lambda^0 \in \operatorname{int} \Longrightarrow \Gamma$ locally Lipschitzian at $\lambda^0$.

Proof. See KLATTE/KUMMER (1987).

Theorem 4. permits to state that the constraint set mapping M given by (5) is locally Lipschitzian on $\operatorname{int}\Lambda$ if

(i) in (5) all functions are convex and (CPC) is satisfied or,

(ii) in (5) all functions are primitive convex polynomial functions (for the definition and properties of such polynomial functions see 4.3.).

However, the following simple example (see KLATTE (1985)) illustrates that a locally Lipschitzian behaviour can even not be guaranteed for the optimal set function $\psi$ of an optimization problem corresponding to the classes (i) and (ii) above:

$$\inf\{x_1 \ / \ -x_1 + x_2^2 \leq 0,\ -x_1 \leq \lambda\},\quad \lambda \in R,$$

$$\psi(\lambda) = \{(0,0)\} \quad \forall \lambda \geq 0,$$

$$\psi(\lambda) = \{x \in R^2 \ / \ x_1 = -\lambda,\ -\sqrt{-\lambda} \leq x_2 \leq \sqrt{-\lambda}\} \quad \forall \lambda < 0.$$

The optimal set function is not locally upper Lipschitzian at $\lambda = 0$ ($\psi$ is not a convex multifunction).

In this monograph we do not deal any more with the locally Lipschitzian behaviour of the mappings related to parametric optimization problems. For a current survey the reader is referred to the monograph KLATTE/KUMMER (1987).

Now, we shall derive some tools which will be used in Chapter 5., where we investigate constraint sets of the form (5) and, additionally, integer requirements are imposed on variables. For this purpose we assign to the multifunction M defined by (5) another multifunction

$M_{st} : \Lambda \longrightarrow 2^{R^n}$ where $\Lambda$ = dom M

given by

$$M_{st}(\lambda) \underset{Df}{=} \{x \in R^n \ / \ p_j(x) \leq \lambda_j, \ j \in J_{st}\} \tag{21}$$

and

$$J_{st} \underset{Df}{=} \{ j \in \{1,\dots,m\} / \inf\{p_j(x) \ / \ x \in M(\lambda)\} > -\infty\}. \tag{22}$$

For simplicity we consider the mapping $M_{st}$ on the set $\Lambda$ although, in general, not all parameters $\lambda_j$ appear in (21).
For a fixed $\lambda$ the set $M_{st}(\lambda)$ is called the stable set to $M(\lambda)$ and the corresponding multifunction $M_{st}$ is called the stable mapping to M. We note that according to Theorem 2. the index set $J_{st}$ (22) is actually independent of $\lambda \in \Lambda$.
In order to characterize the stable stets $M_{st}(\lambda), \lambda \in \Lambda$ (or, more precisely, to determine the index set $J_{st}$) we apply the following elimination procedure for systems of quasiconvex polynomial constraints which describe the sets $M(\lambda)$.

The elimination procedure.

Step 0. Set $J_0 := \{1,\dots,m\}$.

Step r. If for a $j_r \in J_{r-1}$ there is a direction
$u_{j_r} \in 0^+\{x \in R^n \ / \ p_j(x) \leq \lambda_j, \ j \in J_{r-1}\}$
with $u_{j_r} \notin R(p_{j_r})$, then set

$J_r := J_{r-1} \setminus \{j_r\}$

and go to Step (r+1).;
otherwise, set $J^* := J_{r-1}$ and stop.

Note, that the considered direction $u_{j_r}$ in Step r. exists if and only if the set $\{x \in R^n \ / \ p_j(x) \leq \lambda_j, \ j \in J_{r-1}\}$ contains a half-line $\{x \ , \ x = x^0 + \alpha u_{j_r}, \alpha \geq 0\}$ on which the polynomial function $p_{j_r}(x)$ is unbounded from below.

Lemma 9.
Let $M(\lambda), \lambda \in \Lambda$, be given by (5). The index set $J^*$ generated by the above elimination procedure coincides with the index set $J_{st}$ of the stable sets $M_{st}(\lambda)$.

Proof.
Because of the stopping criteria the set $J^*$ contains only indices j

such that the polynomial functions $p_j(x)$ are constant on the recession cone $0^+M_{J^*}(\lambda)$ of the sets

$$M_{J^*}(\lambda) \underset{Df}{=} \{x \in R^n \ / \ p_j(x) \leq \lambda_j, \ j \in J^*\} .$$

Hence the recession cone $0^+M_{J^*}(\lambda)$ is a linear subspace and the closed convex set $M_{J^*}(\lambda)$ may be represented as the sum of a compact set and the linear subspace $0^+M_{J^*}(\lambda)$. Therefore, all polynomial functions $p_j(x)$ $j \in J^*$, are bounded from below on $M_{J^*}(\lambda)$ and, with $M(\lambda) \subset M_{J^*}(\lambda)$ it follows that $J^* \subset J_{st}$.

In order to show the opposite inclusion we assume the following situation after the execution of Step r.:

$$J_{r-1} \subset J_{st}, \text{ but } j_r \notin J_{st}.$$

Then we have

$$c \underset{Df}{=} \inf\{p_{j_r}(x) \ / \ x \in M(\lambda)\} > -\infty$$

and

$$\inf\{p_{j_r}(x) \ / \ p_j(x) \leq \lambda_j, \ j \in J_{r-1}\} = -\infty.$$

According to the choice of $u_{j_{r-1}}, \ldots, u_{j_1}$ and Theorem 2. we successively obtain

$$\{x \in R^n \ / \ p_j(x) \leq \lambda_j, \ j \in J_{r-2}, \ p_{j_r}(x) \leq c-1\} \neq \emptyset,$$

..........................................

$$\{x \in R^n \ / \ p_j(x) \leq \lambda_j, \ j \in J_0, \ p_{j_r}(x) \leq c-1\} \neq \emptyset,$$

which contradicts the definition of c.

Therefore, $J^* \supset J_{st}$ follows. #

Corollary 4.

(i) The recession cone

$$V_{J_{st}} \underset{Df}{=} 0^+M_{st}(\lambda)$$

of the stable sets $M_{st}(\lambda)$, $\lambda \in \Lambda$, is a linear subspace and the multifunction $M_{st}$ is continuous on $\Lambda$.

(ii) The domain dom $M_{st}$ coincides with $\Lambda =$ dom M and it is a closed set.

Proof. By Lemma 9. and its proof, Remark 5. and Corollary 1. #

The proposition (i) of Corollary 4. justifies the notations of a stable set and a stable mapping.

The following lemma is strongly related to the consideration of integer points in constraint sets of the form (5).

Lemma 10.

Let $\lambda \in \Lambda$ and $M_{st}(\lambda)$ be the stable set to $M(\lambda)$ (5). Then, for every $\varepsilon > 0$, there exists a point $x^* \in M_{st}(\lambda)$ such that

$$p_j(x) \leq \lambda_j \quad \forall\, x \in U_\varepsilon \{x^*\} \quad \forall\, j \notin J_{st} \tag{23}$$

holds.

Proof.

We apply the above elimination procedure to the system describing $M(\lambda)$. By $u^{j_1},\dots,u^{j_r}$ we denote the direction considered in the single steps. Let $x^0$ be an arbitrarily chosen point of $M_{st}(\lambda)$. Since the quasiconvex polynomial function $p_{j_r}(x)$ decreases over each half-line of the form

$$S = \{x \in R^n \;/\; x = y + t\, u^{j_r},\ t \geq 0\},\ y \in R^n,$$

we can choose a number $t_r > 0$, sufficiently large, such that

$$p_{j_r}(x) \leq \lambda_{j_r} \quad \forall x \in \mathrm{cl}\ U_\varepsilon\{x^0 + t_r u^{j_r}\}.$$

Furthermore, the point $x^0 + t_r u^{j_r}$ belongs to $M_{st}(\lambda)$ since

$$p_j(x^0 + t_r u^{j_r}) = p_j(x^0) \leq \lambda_j \quad \forall\, j \in J_{st}.$$

Now we choose a number $t_{r-1} > 0$ sufficiently large such that

$$p_{j_{r-1}}(x) \leq \lambda_{j_{r-1}} \quad \forall x \in \mathrm{cl}\ U_\varepsilon\{x^0 + t_r u^{j_r} + t_{r-1} u^{j_{r-1}}\}$$

holds. From

$$p_j(x^0 + t_r u^{j_r} + t_{r-1} u^{j_{r-1}}) = p_j(x^0 + t_r u^{j_r}) \leq \lambda_j \quad \forall j \in J_{st}$$

and

$$p_{j_r}(x) \leq \sup\{p_{j_r}(x) \;/\; x \in \mathrm{cl}\ U_\varepsilon\{x^0 + t_r u^{j_r}\}\} \leq \lambda_{j_r}$$
$$\forall\, x \in \mathrm{cl}\ U_\varepsilon\{x^0 + t_r u^{j_r} + t_{r-1} u^{j_{r-1}}\}$$

one concludes

$$x^0 + t_r u^{j_r} + t_{r-1} u^{j_{r-1}} \in M_{st}(\lambda),$$

$$p_{j_r}(x) \leq \lambda_{j_r},\ p_{j_{r-1}}(x) \leq \lambda_{j_{r-1}} \quad \forall\, x \in \mathrm{cl}\ U_\varepsilon\{x^0 + t_r u^{j_r} + t_{r-1} u^{j_{r-1}}\}.$$

Proceeding in this way we obtain the point

$$x^* = x^0 + t_r u^{j_r} + \dots + t_1 u^{j_1} \in M_{st}(\lambda)$$

fulfilling

$$p_{j_r}(x) \leq \lambda_{j_r}, \dots, p_{j_1}(x) \leq \lambda_{j_1} \quad \forall x \in \mathrm{cl}\ U_\varepsilon\{x^*\}. \quad \#$$

## 4.3. Stability Properties for a Fixed Objective Function

We are looking for the stability of the following parametric optimization problem:

$$P(\lambda) \quad \inf\{p(x) \;/\; x \in M(\lambda)\},\ \lambda \in \Lambda = \mathrm{dom}\ M,$$

where $M(\lambda)$ is given by (5) and $p: R^n \longrightarrow R$ is a quasiconvex polynomial function.

By our considerations above we know that:

1.) M is a l.s.c. mapping on $\Lambda$ (Theorem 1.),

2.) M is H-continuous on $\Lambda$ if the sets $M(\lambda)$ satisfy (CPC) (Corollary 1.)

3.) If the condition

$$\exists \lambda^{o} \in \Lambda \ : \varphi(\lambda^{o}) > -\infty \tag{24}$$

is satisfied, then the value function $\varphi$ given by (8) is continuous on $\Lambda$ ; otherwise, $\varphi(\lambda) = -\infty$ for all $\lambda \in \Lambda$ (Theorem 2.).

From Theorem 7.4. one easily deduces for $P(\lambda)$ that:

4.) The optimal sets $\psi(\lambda)$, $\lambda \in \Lambda$ are non-empty if (24) is satisfied.

As a consequence of Corollary 1. we obtain

Theorem 5.

Let $M(\lambda)$ be given by (5) and p be a quasiconvex polynomial function. Further, let (24) hold. If the set

$$\hat{M} \underset{Df}{=} \{x \in R^{n} \ / \ p(x) \leq p(0), \ p_i(x) \leq p_i(0), \ i = 1,\ldots,m\}$$

satisfies the condition (CPC), then there is a compact-valued mapping $\psi^{*} : \Lambda \longrightarrow 2^{R^{n}}$ which is H-continuous on $\Lambda$ such that

$$\psi(\lambda) = \psi^{*}(\lambda) + 0^{+}\hat{M} \quad \forall \lambda \in \Lambda.$$

Proof. One can write, for each $\lambda \in \Lambda$,

$$\psi(\lambda) = M(\lambda) \cap \{x \in R^{n} \ / \ p(x) \leq \varphi(\lambda)\}.$$

By assumption, (CPC) is satisfied for $\hat{M}$, which implies that (CPC) is also fulfilled for every set $\psi(\lambda)$, $\lambda \in \Lambda$ . Taking into account that $\varphi$ is continuous the proposition follows by Theorem 3. and Corollary 1. #

In particular, the condition (CPC) is satisfied for the set $\hat{M}$ if it holds for the constraint sets $M(\lambda)$, and (24) is ensured. This becomes clear if one respects that the optimal sets may be represented by

$$\psi(\lambda) = M(\lambda) \cap \{x \in R^{n} \ / \ x = x^{*} + R(p)\} \ , \lambda \in \Lambda \quad \text{fixed},$$

where $x^{*} \in \psi(\lambda)$ is fixed and $R(p)$ denotes the linear space of constancy of p as used in Lemma 4.

In BANK/BELOUSOV et al. (1986) SHIRONIN exposed a class of so-called primitive convex polynomial functions which shows the property that if the $p_i$ appearing in $M(\lambda)$ given by (5) are taken from it, then the mapping M is H-continuous on $\Lambda$ . A convex polynomial function $f : R^{n} \longrightarrow R$ is said to be primitive if the following condition holds: If any two affine subspaces $L_1$ and $L_2$ from the space $R^{n}$ satisfy

$$L_1 \parallel L_2, \ L_2 \parallel L_1, \quad L_i \cap \text{int epi } f \neq \emptyset, \ i = 1,2,$$

$$\Longrightarrow (\text{epi } f \cap L_1) \mid (\text{epi } f \cap L_2)$$

(Here, $M_1 \parallel M_2$ means "$M_1$ is parallel to $M_2$", i.e. $\sup_{x \in M_1} d(x,M_2) < \infty$).

Due to SHIRONIN (1980) we have:

(i) The primitivity of a convex polynomial function is invariant with respect to a linear transformation.

(ii) The sum of finitely many primitive convex polynomial functions yields a primitive polynomial function.

(iii) Convex forms are primitive.

(iv) A convex polynomial function is primitive if it is non-linear in at most two variables.

(v) A convex polynomial function f on $R^3$ is primitive if its degree does not exceed 14.

As a consequence of his considerations on the H-continuity of $M(\lambda)$ (5) involving only primitive convex polynomial function SHIRONIN obtained the following theorem:

<u>Theorem 6.</u>
If all polynomial functions appearing in $P(\lambda)$ are convex and $M(\lambda)$ is defined by primitive polynomial functions, then the optimal set-function $\psi$ is H-continuous on $\Lambda$ if (24) is satisfied.

This theorem remains valid if the objective function p is only a quasi-convex polynomial function.

## 5. Integer Points in Certain Subsets of the Space $R^n$

This chapter is devoted to properties of the constraint sets G(b) of (parametric) mixed-integer optimization problems. In particular, such properties of sets of (mixed-) integer points G(b) will be considered which are of importance for the analysis of the problems studied in the later chapters.

Mixed-integer points in sets in $R^n$ which are representable as the vector sum of a bounded set and a convex cone can be characterized in view of their "distribution"; this is carried out in Section 5.1. For sets M(b) of the special form (2.15) we can show that the mappings G and $G_{conv}$ defined by

$$G(b) = \{x \in M(b) \;/\; x_1,\dots,x_s \text{ integer}\}$$

and

$$G_{conv}(b) = \text{conv } G(b)$$

are u.s.c. if the included polyhedral cone V has a system of (mixed-) integer generators (i.e. V satisfies the (MIG) assumption; see below).

In Section 5.2. we consider the sets G(b) of the (mixed-) integer points in sets M(b) described by quasiconvex polynomial inequalities (4.5). If, in particular, M(b) belongs to the class of sets defined by (2.15) (this is the case in Theorem 4.3.), then, under the (MIG)-assumption, we shall expose subsets of dom G on which the mapping G is even continuous.

For linearly described sets M(b) one can derive more extended properties of G(b) and the related mappings. This is pointed out in Section 5.3.; here we shall also show a "quasi-Lipschitzian" behaviour of the constraint set-mapping G and characterize the vertices of the convex hull conv G(b).

### 5.1. Subsets Described as the Sum of a Compact Set and a Convex Cone

If M is a subset of $R^n$ and s, $0 \leq s \leq n$, a natural number, then we denote the *set of (mixed-) integer points* belonging to M by

$$M_s \underset{Df}{=} \{x \in M \;/\; x_1,\dots,x_s \text{ integer}\}.$$

We (analogously to BELOUSOV (1977)) say that the points of $M_s$ are *uniformly distributed* in the set $M \subset R^n$ if

$$M_s \neq \emptyset$$

and

$$\sup\{d(x,M_s) \;/\; x \in M\} < \infty.$$

The latter condition means that the ball around every point of M having this supremum as radius contains at least one point of $M_s$.

From Diophantine Approximation we cite the following lemma, which is related to KRONECKER's famous Approximation-Theorem:

<u>Lemma 1.</u>
If $k \in R^n$ is a non-zero vector, then, for all $\varepsilon > 0$, the integer points of the set

$$S_\varepsilon = \{x \in R^n \;/\; x = tk + y,\ \|y\| < \varepsilon\ ,\ t \in R\}$$

are uniformly distributed in $S_\varepsilon$ .

<u>Proof.</u> See for example BELOUSOV (1977). #

<u>Corollary 1.</u>
If L is a linear subspace, then, for all $\varepsilon > 0$, the integer points of the set

$$L = \{x \in R^n \;/\; x = z + y,\ z \in L,\ \|y\| < \varepsilon\} \tag{1}$$

are uniformly distributed in $L_\varepsilon$ .

<u>Proof.</u> Let dim L = q $\geq$ 1. Then one can write

$$L = s_1 + \ldots + s_q$$

where $\{s_1, \ldots, s_q\}$ is the system of the 1-dimensional orthogonal subspaces of L. For a fixed $\varepsilon > 0$ we find an $\varepsilon^* > 0$ such that the set

$$L^* \underset{Df}{=} \{x \in R^n \;/\; x = x^1 + y^1 + \ldots + x^q + y^q,\ x^i \in s_i,\ \|y^i\| < \varepsilon^*,\ i = 1, \ldots, q\}$$

is a subset of $L_\varepsilon$ . It is sufficient to show that the integer points of the set $L^*$ are uniformly distributed in $L^*$. However, this is clear since by Lemma 1. the integer points of every set

$$S^i_{\varepsilon^*} = \{x \in R^n \;/\; x = x^i + y^i,\ x^i \in s_i,\ \|y^i\| < \varepsilon^*\}$$

are uniformly distributed. #

<u>Lemma 2.</u>
If $V \subset R^n$ is a convex cone, then, for all $\varepsilon > 0$, the integer points in the set

$$V_\varepsilon \underset{Df}{=} \{x \in R^n \;/\; x = z + y,\ z \in V, \|y\| < \varepsilon\}$$

are uniformly distributed in $V_\varepsilon$ .

<u>Proof.</u> The proposition has only to be shown for $V \neq \{0\}$. If we denote by L the linear hull of V and arbitrarily fix an $\varepsilon > 0$, then, from Corollary 1., it follows that the integer points of the set $L_\varepsilon$(1) are uniformly distributed in $L_\varepsilon$. Let $\delta$ be such that

$$\delta > \sup\{d(x,(L_\varepsilon)_n) \;/\; x\in L_\varepsilon\}$$

holds, where

$$(L_\varepsilon)_n \underset{Df}{=} \{x\in L_\varepsilon / \; x_1,\ldots,x_n \text{ integer}\}.$$

Now there is a point $x^*\in V$ with

$$U_\delta\;(x^*)\cap L\subset V.$$

Let x be an arbitrary point of $V_\varepsilon$. Then there are points $z\in V$ and $y\in R^n$ with $\|y\|<\varepsilon$ such that

$$x = z + y.$$

Then $x^* + z\in V$ and

$$U_\delta(x^*+z)\cap L\subset V,$$

from which it follows that

$$A \underset{Df}{=} (U_\delta(x^*+z)\cap L) + U_\varepsilon(0)\subset V_\varepsilon .$$

Now let

$$a\in U_\delta(x^*+z)\cap L_\varepsilon \qquad (2)$$

be an integer point existing by the choice of $\delta$ (and Corollary 1.). Then there is a point $u\in R^n$ with $\|u\|<\delta$ such that

$$a = x^* + z + u,$$

and we have

$$\begin{aligned}\|x-a\| &= \|z+y - x^*-z-u\| = \\ &= \|y-x^*-u\|\le \|y\| + \|x^*\| + \|u\| < \\ &< \|x^*\| + \varepsilon + \delta ,\end{aligned}$$

which gives us an upper bound independent of x. It remains to show that the integer point a belongs to $V_\varepsilon$. But, to see this it is sufficient to verify that $a\in A$.

From (2) we see that the point a may be represented in the form

$$a = a^1 + a^2$$

with

$$a^1\in L,\; a^2\in U_\varepsilon(0) \text{ and } \|a^1+a^2-(x^*+z)\|<\delta.$$

If we decompose the vector $a^2$ as follows:

$$a^2 = x^1 + x^2 \text{ with } x^1\in L,\; x^2\in L^\perp,$$

then we have

$$a^1 + x^1\in L,\; x^* + z\in L \text{ and } x^2\in L^\perp.$$

Therefore

$$\delta > \|a^1 + x^1+x^2-x^*-z\| \ge \|a^1+x^1-x^*-z\|,$$

$$\varepsilon > \|a^2\| = \|x^1+x^2\| \ge \|x^2\|$$

which implies

$$a^1 + x^1 \in U_\delta(x^* + z) \cap L, \quad x^2 \in U_\varepsilon(0),$$

hence $a \in A$. #

For more general considerations we need the following lemma:

Lemma 3.

Let $M \subset R^n$ be an affine subspace with $M_s \neq \emptyset$. The points of $M_s$ are uniformly distributed in M if and only if $M_s$ contains an affine basis of M.

Proof. Assume that $M_s$ contains an affine basis of M, i.e. there are points

$$k^i \in M_s, \quad i = 0, \dots r = \dim M,$$

such that every $x \in M$ may be represented by

$$x = k^0 + \sum_{i=1}^{r} \alpha_i (k^i - k^0)$$

with $\alpha_i \in R$, $i = 1, \dots, r$.

If we consider

$$x' = k^0 + \sum_{i=1}^{r} [\alpha_i](k^i - k^0),$$

then

$$x' \in M_s$$

and

$$\|x - x'\| = \|\sum_{i=1}^{r} (\alpha_i - [\alpha_i])(k^i - k^0)\| \leq \sum_{i=1}^{r} \|k^i - k^0\|.$$

In order to show the opposite direction it suffices to prove aff $M_s \supset M$. Assume that there is a point $x^* \in M \setminus \text{aff } M_s$. Let $x^0 \in \text{aff } M_s$ be fixed, then we have

$$x^0 + t(x^* - x^0) \in M \quad \forall t \in R,$$

$$d(x^0 + t(x^* - x^0), \text{aff } M_s) = |t| \, d(x^*, \text{aff } M_s).$$

But this yields a contradiction to $\sup\{d(x, M_s) \;/\; x \in M\} < \infty$ because

$$d(x^0 + t(x^* - x^0)) \longrightarrow \infty \text{ as } t \longrightarrow \infty. \quad \#$$

Corollary 2.

If M is a linear subspace, then the set $M_n$ of integer points in M is uniformly distributed in M if and only if the orthogonal complement $M^\perp$ has an integer basis.

Proof. Using CRAMER's rule the assertion is easy to see. #

Corollary 3.

Let M be an affine subspace. Then $M_\varepsilon$ coincides with the (mixed-) integer points of aff $M_s$.

Proof. By Lemma 2. #

Theorem 1.
Let $M \subset R^n$ be a set of the form

$$M = C + V$$

where $C \subset R^n$ is a bounded set and $V \subset R^n$ a convex cone. Then the existence of a relatively interior (mixed-) integer point implies that the points of $M_s$ are uniformly distributed in M if the (mixed-) integer points of aff M are uniformly distributed in aff M.

Proof. Without loss of generality we may assume that $0 \in \text{ri } M$.

Let $a^1, \ldots, a^d \in M_s$ be a basis of the linear subspace aff M and, we define

$$\tilde{M} \underset{Df}{=} \{y \in R^d \,/\, \exists\, x \in M : x = \sum_{i=1}^{d} a^i y_i\}.$$

The set $\tilde{M}$ can be represented by

$$\tilde{M} = \tilde{C} + \tilde{V}$$

where $\tilde{C}$ is a bounded subset of $R^d$ and $\tilde{V}$ is a convex cone in $R^d$. Obviously, $0 \in \text{int } \tilde{M} \subset R^d$. By Lemma 2., for all $\varepsilon > 0$, the integer points of

$$\tilde{V}_\varepsilon = \{y \in R^d \,/\, y = z + u,\ z \in \tilde{V},\ \|u\| < \varepsilon\}$$

are uniformly distributed in $\tilde{V}_\varepsilon$ .
Since $\tilde{C}$ is a bounded set and $0 \in \text{int } \tilde{M}$, we may conclude that the integer points of $\tilde{M}$ are also uniformly distributed in $\tilde{M}$, i.e.

$$\delta = \sup \{d(x, (\tilde{M})_d) \,/\, x \in \tilde{M}\} < \infty .$$

Now let $x \in M$ be an arbitrary point.
Then there are a $y \in \tilde{M}$ and an integer point $y^* \in (\tilde{M})_d$ such that

$$x = \sum_{i=1}^{d} a^i y_i \text{ and } \|y - y^*\| \leq \delta .$$

For the point $x^* = \sum_{i=1}^{d} a^i y_i^* \in M_s$ we have

$$\|x - x^*\| = \|\sum a^i (y_i - y_i^*)\| \leq \delta \sum_{i=1}^{d} \|a^i\|$$

and, therefore

$$\sup\{d(x, M_s) \,/\, x \in M\} < \infty. \quad \#$$

The main part of our considerations will be related to sets represented by the sum of a compact set and a polyhedral cone. If the cone V in Theorem 1. is such a polyhedral cone, then the assumptions may be replaced by requirements on the generators of this polyhedral cone. We use such requirements in the following form:

Let $V \subset R^n$ be a polyhedral cone. We say that V satisfies the assumption (IG) or (MIG), respectively, if

(IG) V has a generating system
$\{v^1,\dots,v^r\} \subset \{v \in R^n \,/\, v_1,\dots,v_n \text{ integer}\}$ and,

(MIG) V has a generating system
$\{v^1,\dots,v^r\} \subset \{v \in R^n \,/\, v_1,\dots,v_s \text{ integer}\}$.

We consider the following set G(b) of (mixed-) integer points

$$G(b) \underset{Df}{=} \{x \in M(b) \,/\, x_1,\dots,x_s \text{ integer}\} \tag{3}$$

where M(b) is given by (2.15), i.e. $M(b) = C(b) \dot{+} V$ and $C : \Lambda \longrightarrow 2^{R^n}$ is a compact-valued u.s.c. multifunction on $\Lambda \subset R^m$ and V a polyhedral cone in $R^n$ (which contains $0 \in R^n$).

The domain of the corresponding multifunction G is given by

$$B \underset{Df}{=} \{b \in \Lambda \,/\, G(b) \neq \emptyset\}. \tag{4}$$

<u>Lemma 4.</u>
Let G(b) be given by (3) and V satisfy (MIG).
Further, let

$$\delta(b) \underset{Df}{=} \sup\{d(x,G(b)) \,/\, x \in M(b)\}.$$

Then

(i) $\delta(b) < \infty \quad \forall b \in B$,

(ii) $\left. \begin{array}{l} M(b) \cap \{x \in R^n \,/\, x_i \geq \delta(b),\ i=1,\dots,n\} \neq \emptyset, \\ G(b) \neq \emptyset \end{array} \right\}$

$\Longrightarrow G(b) \cap \{x \in R^n \,/\, x \geq 0\} \neq \emptyset$,

(iii) If there is a number $\rho$ such that

$\rho \geq \sup\{d(k,y) \,/\, k,y \in C(b)\} \quad \forall b \in B$,

then there is a $\delta \geq \delta(b)\ \forall b \in B$.

<u>Proof.</u> (i) Let $b \in B$ and $v^1,\dots,v^r$ be generators of V existing by (MIG). Then there is a point

$$x^0 = k^0 + \sum_{i=1}^{r} \alpha_i^0 v^i \in G(b)$$

where $k^0 \in C(b)$ and $\alpha_i^0 \in [0,1)$, $i=1,\dots,r$.
An arbitrarily chosen point $x \in M(b)$ possesses the description

$$x = k + \sum_{i=1}^{r} \alpha_i v^i$$

with $k \in C(b)$ and $\alpha_i \geq 0$, $i=1,\dots,r$.

Setting $\rho(b) \underset{Df}{=} \sup\{d(k^0,y) \,/\, y \in C(b)\}$ and $x' = x^0 + \sum_{i=1}^{r} [\alpha_i] v^i \in G(b)$ one obtains

$$\|x-x'\| = \|k-k^0 + \sum_{i=1}^{r} (\alpha_i - [\alpha_i] - \alpha_i^b)v^i\| \leq$$

$$\leq \varrho(b) + \sum_{i=1}^{r} \|v^i\| \underset{Df}{=} \delta(b).$$

(ii) Let $x^* \in M(b)$ with $x_i^* \geq \delta(b)$, $i=1,\ldots,n$.
Because of the definition of $\delta(b)$ there exists a (mixed-) integer point $x \in G(b)$ with $\|x-x^*\| \leq \delta(b)$, and by the choice of $x_i^* \geq \delta(b)$, $i=1,\ldots,n$, it follows that $x \geq 0$ must hold.

(iii) One estimates

$$\delta(b) = \varrho(b) + \sum_{i=1}^{r} \|v^i\| \leq \varrho + \sum_{i=1}^{r} \|v^i\| \underset{Df}{=} \delta . \quad \#$$

The following example shows that the assumption (IG) in Lemma 4. (i) is not a necessary condition for the uniform distribution of integer points.

Example 1.

The integer points of the set

$$M = \{x \in R^2 \ / \ -1 \leq \sqrt{2}\,x_1 - x_2 \leq 1\}$$

are, by Theorem 1., uniformly distributed in M. Obviously, the set M may be written in the form

$$M = C + V$$

with the bounded set

$$C = \{x \in R^2 \ / \ x_1 = 0, \ -1 \leq x_2 \leq 1\}$$

and the polyhedral cone

$$V = \{u \in R^2 \ / \sqrt{2}\,u_1 - u_2 = 0\}$$

which does not satisfy (IG).
We assign to the sets G(b) their convex hulls conv G(b) which define the multifunction $G_{conv} : B \longrightarrow 2^{R^n}$ with $G_{conv}(b) \underset{Df}{=} \text{conv } G(b)$.
The next theorem gives a characterization of the mappings G and $G_{conv}$.

Theorem 2.

Let G be given by (3) and let the polyhedral cone V fulfil the condition (MIG). Then we have the following statements:

(i) There exist two compact-valued u.s.c. multifunctions

$$K: B \longrightarrow 2^{R^n} \text{ and } K_{conv}: B \longrightarrow 2^{R^n}$$

such that, for all $b \in B$,

$$G(b) = K(b) + W \text{ and } G_{conv}(b) = K_{conv}(b) + V$$

where either

$$W = V_s \underset{Df}{=} \{u \in V \ / \ u_1,\ldots,u_s \text{ integer}\}$$

or

$$W = \{u \in R^n \;/\; u = \sum_{i=1}^{r} \beta_i v^i, \beta_i \text{ natural}, \; i=1,\dots,r\}$$

if $v^1,\dots,v^r$ are (mixed-) integer generators of V;

(ii) The multifunctions G and $G_{conv}$ are u.s.c. on B;

(iii) The domain B is closed if $\Lambda$ is closed.

Proof.

Let us define the multifunctions K and $K_{conv}$ by

$$K(b) \underset{Df}{=} \{x \in R^n \;/\; x = y + \sum_{i=1}^{r} \alpha_i v^i, \; 0 \le \alpha_i \le 1, \; i=1,\dots,r; \; y \in C(b), \; x_1,\dots,x_s \text{ integer}\} \tag{5}$$

and

$$K_{conv}(b) \underset{Df}{=} \text{conv } K(b), \tag{6}$$

respectively, which are given on B. The appearing vectors $v^i$, $i=1,\dots,r$, are (mixed-) integer generators of the cone V. Using now the u.s.c. property of C and the fact that $Z^s \times R^{n-s}$ is a closed set (here, Z denotes all integer numbers) one obtains the claimed upper semicontinuity of K, of which the same property of $K_{conv}$ is an immediate consequence. The desired representation G(b) = K(b) + W is in both cases obvious. For an arbitrary $b \in B$ one has

$$\begin{aligned} \text{conv } G(b) &= \text{conv}(K(b)+W) = \\ &= \text{conv } K(b) + \text{conv } W = \\ &= K_{conv}(b) + V, \end{aligned}$$

which completes the proof of (i).

The upper semicontinuity of the multifunctions G and $G_{conv}$ is now a consequence of the u.s.c. property of K and $K_{conv}$.

In order to show (iii) we consider an arbitrary sequence $\{b^t\} \subset B$ with $b^t \longrightarrow b^o$. Since

$$K(b) \neq \emptyset \iff G(b) \neq \emptyset,$$

we have $K(b^t) \neq \emptyset$ for t = 1,2,... . By $C(b^o) \neq \emptyset$ and the construction (5) of K(b) it follows that $K(b^o) \neq \emptyset$, i.e. $b^o \in B$. #

In order to guarantee the continuity of the constraint set mapping G it does not suffice to impose the stronger hypothesis that C in (3) is a continuous multifunction on $\Lambda$. This is shown by the following simple example

$$G(b) = \{x \in R \;/\; 0 \le x \le b, \; x \text{ integer}\}.$$

We have

$$G(b_t) = \{0\} \text{ and } G(b_o) = \{0,1\}$$

if we consider

$$b_t \underset{Df}{=} (1 - \tfrac{1}{t}) \longrightarrow b_o = 1.$$

However, we shall show in the following section that for a constraint set mapping G, where the included M is given by the set of solutions of a system of quasiconvex polynomial inequalities with perturbed right-hand sides, it is possible (of course under certain assumptions) to characterize subsets of B on which the restriction of G is continuous.

## 5.2. Subsets Described by Quasiconvex Polynomial Functions

This section is devoted to sets of (mixed-) integer points given by

$$G(b) \underset{Df}{=} \{x \in M(b) \ / \ x_1,\ldots,x_s \text{ integer}\} \tag{7}$$

where M(b) is given by (4.5), i.e.

$M(b) = \{x \in R^n \ / \ p_j(x) \leq b_j, \ j=1,\ldots,m\}$

and $p_j : R^n \longrightarrow R$, $j = 1,\ldots,m$, are quasiconvex polynomial functions.

By $\Lambda$ and B we denote the domains of the corresponding mappings M and G, respectively.

First, we derive conditions under which a quasiconvex polynomial system of inequalities has a (mixed-) integer solution.

<u>Lemma 5.</u>
If J is an index set fulfilling $J_{st} \subset J \subset \{1,\ldots,m\}$, where $J_{st}$ is given by (4.22), then the set G(b) is non-empty if the (mixed-) integer points in the set

$$M_J(b) \underset{Df}{=} \{x \in R^n \ / \ p_j(x) \leq b_j, \ j \in J\} \tag{8}$$

are uniformly distributed in $M_J(b)$.

<u>Proof.</u> Let

$\delta > \sup\{d(x,(M_J(b))_s) \ / \ x \in M_J(b)\}$.

Then, by Lemma 4.10., there is a point $x^* \in M_{st}(b)$ such that

$p_j(x) \leq b_j \ \forall j \notin J_{st} \ \ \forall x \in U_\delta(x^*)$

holds. Because of the choice of $\delta$ there exists an $\tilde{x} \in (M_J(b))_s$ satisfying $\|x^*-\tilde{x}\| < \delta$ and we obtain $\tilde{x} \in G(b)$. #

If one requires the (MIG) assumption for the recession cone $V_{J_{st}}$ of the stable sets $M_{st}(b)$, then the (mixed-) integer points of $M_{st}(b)$ are uniformly distributed in $M_{st}(b)$ whenever the latter set contains a (mixed-) integer point.

<u>Theorem 3.</u>
Let the recession cone $V_{J_{st}}$ of the stable set $M_{st}(b)$ fulfil the assumption

(MIG). Then

(i) $G(b) \neq \emptyset \Longleftrightarrow (M_{st}(b))_s \neq \emptyset$;

(ii) the domain B of the multifunction G is closed.

Proof. (i) By Corollary 4.4.(i) the stable sets $M_{st}(b)$ fulfil the condition (CPC), and now the (MIG) assumption for $V_{J_{st}}$ yields the implication

$$(M_{st}(b))_s \neq \emptyset \Longrightarrow G(b) \neq \emptyset,$$

if one takes Lemma 4. and Lemma 5. into account. The opposite implication is obvious.

(ii) By Corollary 4.4.(ii) the domain of the stable mapping $M_{st}$ defined by (4.21) is closed. Using now Theorem 2. one obtains that the set $\{b \in R^m \,/\, (M_{st}(b))_s \neq \emptyset\}$ is closed and, applying (i), the proposition (ii) follows. #

The following example shows that the (MIG) assumption for $V_{J_{st}}$ is essential for the validity of Theorem 5.

Example 2.

The stable sets $M_{st}(b)$ of the sets

$$M(b) = \{x \in R^2 \,/\, \sqrt{2}\, x_1 - x_2 \leq b,\ -\sqrt{2}\, x_1 + x_2 \leq 0,\ -x_1 \leq -1\}$$

are given by

$$M_{st}(b) = \{x \in R^2 \,/\, \sqrt{2}\, x_1 - x_2 \leq b,\ -\sqrt{2}\, x_1 + x_2 \leq 0\}.$$

The recession cone $V_{J_{st}}$ of $M_{st}(b)$ is the linear subspace

$$V_{J_{st}} = \{u \in R^2 \,/\, \sqrt{2}\, u_1 - u_2 = 0\}$$

and, it has no integer generator. If s = 2, then $G(b_o) = \emptyset$ for $b_o = 0$. But $(M_{st}(b_o))_2 = \{\binom{0}{0}\}$ and, for the sequence

$b_t = \frac{1}{t} \longrightarrow 0 = b_o$ one has $G(b_t) \neq \emptyset$.

By Remark 4.6. we know from Theorem 2. that the multifunction G (7) is u.s.c. on its domain if one imposes the assumptions (CPC) on the mapping M involved in (7) and (MIG) on the recession cone of M(b). However, in order to be able to guarantee the continuity of this mapping G one has to restrict it to certain subsets $B(b^o)$ of the domain B.

Let $b^o \in B$ be a fixed point. We define

$$B(b^o) \underset{Df}{=} \{b \in B \,/\, P_{x_1,\dots,x_s} G(b) = P_{x_1,\dots,x_s} G(b^o)\} \qquad (8)$$

where $P_{x_1,\dots,x_s}$ denotes the projection into the subspace of $R^n$ corresponding to the integer variables $x_1,\dots,x_s$.

One immediately shows that the following holds for all fixed $b^o \in B$:

$$B(b^o) \neq \emptyset; \qquad (9)$$

$$b^1 \in B(b^o) \Longrightarrow B(b^1) = B(b^o); \qquad (10)$$

$$b^1 \in B \setminus B(b^o) \Longrightarrow B(b^1) \cap B(b^o) = \emptyset. \qquad (11)$$

<u>Theorem 4.</u>

Let (CPC) be satisfied for M in (7) and let (MIG) be fulfilled for the recession cone V. Then

(i) G is u.s.c. on B;

(ii) G is continuous on $B(b^o)$ for an arbitrarily fixed $b^o \in B$.

<u>Proof.</u> (i) is clear by the remark preceeding this theorem.

(ii) Because of (i) we have only to show the l.s.c. property of G on $B(b^o)$. Let

$$b^* \in B(b^o), \ \{b^t\} \subset B(b^o) \text{ and } b^t \longrightarrow b^*.$$

Obviously, it suffices to verify that for each arbitrarily fixed $x^* \in G(b^*)$ there exists a sequence $\{x^t\} \subset R^n$ with $x^t \in G(b^t)$, for almost all t, such that $x^t \longrightarrow x^*$. Hence, let $x^* \in G(b^*)$ be fixed. We consider the multifunction $G(x_1^*,\dots,x_s^*)$ defined on $B(b^o)$ by

$$G(x_1^*,\dots,x_s^*;b) \underset{Df}{=} \{x \in M(b) \ / \ x_j = x_j^*, \ j=1,\dots,s\}. \qquad (12)$$

We have $G(x_1^*,\dots,x_s^*;b^t) \neq \emptyset \ \forall t$ because of the definition of $B(b^o)$ and it holds that $x^* \in G(x_1^*,\dots,x_s^*;b^*)$. By Theorem 2.5. and Theorem 4.3. we conclude that the mapping $G(x_1^*,\dots,x_s^*)$ is continuous on $B(b^o)$. Hence, there is a sequence $\{x^t\} \subset R^n$ such that

$$x^t \in G(x_1^*,\dots,x_s^*;b^t) \ \forall t \text{ and } x^t \longrightarrow x^*.$$

Since $G(x_1^*,\dots,x_s^*;b^t) \subset G(b^t)$, the desired sequence is obtained by $\{x^t\}$.

#

The following example shows that the validity of the proposition (ii) of Theorem 4. essentially depends on the underlying description of the involved multifunction M.

For the compact-valued continuous multifunction $M: [-\frac{1}{4},\frac{1}{4}] \longrightarrow 2^{R^2}$ given by

$$M(b) \underset{Df}{=} \{x \in R^2 / x_2 \leq 1, \ 2x_1 - x_2 \leq 1, \ -2x_1 - x_2 \leq 1, \ (1-b)x_1 + bx_2 = -b\}$$

and the constraint set mapping G by

$$G(b) = \{x \in R^2 \ / \ x \in M(b), \ x_1 \text{ integer}\}$$

we have

$$B \equiv \Lambda = [-\tfrac{1}{4},\tfrac{1}{4}], \ B(0) = B,$$

but, obviously, G is not l.s.c. at $b = 0 \in B(0)$.

Now we cite a result published by TARASOV/KHACHIYAN (1980), which gives an estimate of the norm of at least one point belonging to $G(0) \neq \emptyset$ in the case that all polynomials $p_j(x)$ appearing in (7) are convex with integer coefficients. Therefore, let us denote by $d_j$ the degree of $p_j(x)$ and $d \underset{Df}{=} \max_j d_j$; further, if $d_{i_1} \geq d_{i_2} \geq \ldots \geq d_{i_m}$, then $D \underset{Df}{=} d_{i_1}, \ldots, d_{i_p}$ where $p \underset{Df}{=} \min\{m,n\}$, and finally, let h be the maximum of the moduli of the coefficients in all polynomials appearing in (7).

<u>Theorem 5.</u>

Let all the polynomials $p_j$ in (7) be convex with integer coefficients, and let d, D and h be defined as above. If $d \geq 2$ and G(0) is non-empty, then there is an $x \in G(0)$ fulfilling

$$\|x\| \leq \omega, \text{ where} \qquad (13)$$
$$\log \omega = \frac{1}{2} D \, (d^{3} n)^{\frac{d}{2}} \log(h \cdot dn).$$

Theorem 5. was used by TARASOV/KHACHIYAN in order to discuss the algorithmic complexity for convex polynomial optimization (for a comprehensive survey the reader is referred to the paper of KHACHIYAN (1982)).

In the context of the considerations presented here, the essential difficulty of deriving an estimate of the form (13) consists in obtaining an estimate $r = r(d,D,h,n)$ such that

$$M_{st}(0) \cap V^{\perp}_{J_{st}} \subset \{x \in R^n \;/\; \|x\| \leq r\}$$

holds for the stable set $M_{st}(0)$ of $G(0) \neq \emptyset$. This becomes clearer by the facts that

$$(M_{st}(0))_s \neq \emptyset \Longleftrightarrow G(0) \neq \emptyset$$

(Theorem 3.(i)) and that the recession cones $0^{+}\{x \in R^n / p_j(x) \leq 0, \; j \in J\}$, $J \subset \{1,\ldots,m\}$, can be described applying the techniques of Chapter 4. Using the radius r one can easily obtain a radius $r^* = r^*(d,D,h,n)$ such that

$$\exists \, x^* \in (M_{st}(0))_s \cap \{x \in R^n \;/\; \|x\| \leq r^*\}.$$

Starting with such an $x^*$ and following the way described in the proof of Lemma 4.10. one can construct natural numbers $t_1,\ldots,t_k$ and (mixed-) integer vectors $u^1,\ldots,u^k$ (where $k \leq m-|J_{st}|$) such that

$$x^* + \sum_{i=1}^{k} t_i u^i \in G(0). \qquad (14)$$

In a more precise analysis one can verify that the numbers $t_i$ and vectors $u^i$ in (14) can be chosen in such a way that

$$\|x^* + \sum_{i=1}^{k} t_i u^i\| \leq r^* + \sum_{i=1}^{k} t_i \|u^i\| \leq \rho$$

holds, where $\rho$ has the desired form.

## 5.3. Polyhedral Subsets

In this section we consider (mixed-) integer points in subsets of the form (4.5) where the polynomial functions are linear, i.e.

$$p_j(x) = d^{j^T} x \ , \ j=1,\dots,m.$$

Obviously, then the recession cone V of the sets $M(b) \neq \emptyset$ is given by

$$V = \{u \in R^n \ / \ d^{j^T} u \leq 0, \ j=1,\dots,m\}.$$

By Corollary 4.2. the multifunction G given by

$$G(b) \underset{Df}{=} \{x \in R^n \ / \ d^{j^T} x \leq b_j, \ j=1,\dots,m; \ x_1,\dots,x_s \text{ integer}\} \tag{15}$$

belongs to the class considered in Theorem 2. where the compact sets K(b) from (5) are specified by C(b) in the form (4.17).

Theorem 6.

Let G be given by (15) and, let V satisfy the (MIG)-condition. Further, let K be given by (5) and (4.17), and let $K_{conv}$ and $G_{conv}$ have a corresponding meaning as in Theorem 2. Then

(i) each set K(b) is a union of a finite number of compact convex polyhedra;

(ii) the sets $K_{conv}(b)$ and $G_{conv}(b)$ are convex polyhedra;

(iii) the polyhedron $G_{conv}(b)$ has a vertex if and only if $\{0\}$ is the only subspace contained in V;

(iv) there exist positive numbers $\rho$ and $\tau$ (independent of b) such that $\|x\| \leq \rho\|b\| + \tau \ \forall x \in K(b)$.

Proof.

The statements (i), (ii) and (iii) are simple consequences of Theorem 2. and the well-known representation theorems for convex polyhedra. In order to prove (iv) we use the constant number $\rho > 0$ from Corollary 4.2. corresponding to the multifunction C (4.17). If $v^1,\dots,v^r$ are (mixed-) integer generators of V and if we take $\tau \underset{Df}{=} \sum_{i=1}^{r} \|v^i\|$, then we obtain the desired estimate for the sets K(b) (5). #

We call a multifunction $\Gamma : \Lambda \longrightarrow 2^{R^n}$ quasi-Lipschitzian on a subset $\Lambda_0 \subset \Lambda$ if there are constant numbers $\rho > 0$ and $\tau > 0$ (not depending on $\lambda \in \Lambda$) such that

$$d(\Gamma(\lambda), \Gamma(\lambda')) \leq \rho\|\lambda - \lambda'\| + \tau \quad \forall \lambda, \lambda' \in \Lambda_0.$$

Theorem 7.

Let G(b) (15) be given in the form

$$G(b) = \{x \in R^n \ / \ Dx \leq b, \ x_1,\dots,x_s \text{ integer}\}$$

with a rational (m,n)-matrix D. Then

(i) G is quasi-Lipschitzian on B;

(ii) there exists a positive number $\delta$ such that

$$\delta \geq \sup\{d(x,G(b)) \,/\, x \in M(b)\} \quad \forall b \in B;$$

(iii) there is a compact-valued multifunction $\bar{K} : B \longrightarrow 2^{R^n}$ having the properties:

1° $G(b) = \bar{K}(b) + V_s,\ b \in B,$

2° $\bar{K}$ is quasi-Lipschitzian on B.

<u>Proof.</u> In order to prove (i) one can follow a similar way as in the proof of Corollary 4.3. Let $b, b^* \in B$ and $x \in G(b)$ be arbitrarily chosen. We define

$$\tilde{G}(\beta) \underset{Df}{=} \{(u,v,y,z) \in R^{2(n+m)} \,/\, Du+v-Dy-Dz = \beta,\ v,z \geq 0,\ u_1,\dots,u_s,\ y_1,\dots,y_s \text{ integer}\}$$

and

$$\tilde{V}_{2s} \underset{Df}{=} \{(u,v,y,z) \in R^{2(n+m)} \,/\, Du+v-Dy-z = 0,\ v,z \geq 0,\ u_1,\dots,u_s,\ y_1,\dots,y_s \text{ integer}\}.$$

Then, the Theorems 2. and 6. imply that there exists a multifunction $\tilde{K}$ and constant numbers $\eta > 0$ and $\mu > 0$ such that

$$\tilde{G}(\beta) = \tilde{K}(\beta) + \tilde{V}_{2s},\quad \beta \in \operatorname{dom} \tilde{G},$$

and

$$\|(u,v,y,z)\| \leq \eta\|\beta\| + \mu \quad \forall\, (u,v,y,z) \in \tilde{K}(\beta).$$

Now, by the same arguments as in the proof of Corollary 4.3., one finds a point $x^* \in G(b^*)$ such that

$$\|x-x^*\| \leq 2\eta\|b-b^*\| + 2\mu$$

holds and, setting $\rho = 2\eta$ and $\tau = 2\mu$, (i) is shown.
One obtains proposition (ii) by considering

$$\hat{M} \underset{Df}{=} \{\tbinom{x}{b} \in R^{n+m} \,/\, Dx - b \leq 0\}$$

and

$$\hat{G} \underset{Df}{=} \{\tbinom{x}{b} \in R^{n+m} \,/\, Dx-b \leq 0,\ x_1,\dots,x_s \text{ integer}\}.$$

The points of $\hat{G}$ are uniformly distributed in $\hat{M}$ (Lemma 4.(i)) and we assume

$$\alpha > \sup\{d(x,\hat{G}) \,/\, x \in \hat{M}\}.$$

If $x \in M(b)$, then $\tbinom{x}{b} \in \hat{M}$ and there is a point $\tbinom{x^o}{b^o} \in \hat{G}$ with

$$\|\tbinom{x^o}{b^o} - \tbinom{x}{b}\| < \alpha .$$

Since $x^o \in G(b^o)$, by (i) there is a point $x^* \in G(b)$ such that

$$\|x^o-x^*\| \leq \rho\|b^o-b\| + \tau ,$$

where $\rho$ and $\tau$ are constant numbers independent of b and $b^o$. Now, we can estimate

$$\|x-x^*\| \leq \|x-x^o\| + \|x^o-x^*\| < \alpha + \rho\|b^o-b\| + \tau < \alpha + \rho\alpha + \tau \underset{Df}{=} \delta .$$

(iii) Let $\{v^1,\dots,v^r\}$ be a system of (mixed-) integer generators of

$V = \{x \in R^n \ / \ Dx \leq 0\}$, and let $\tau > \max \{\delta, \sum_{i=1}^{r} \|v^i\|\}$, where $\delta$ is a constant number with the property (ii). Using C(b) defined by (4.17) we define

$$\bar{K}(b) \underset{Df}{=} \{x \in R^n \ / \ x \in G(b),\ x = y+v,\ y \in C(b),\ \|v\| \leq \tau\}.$$

Obviously, $\bar{K}(b)$ is compact and satisfies 1°. In order to show 2° let $b, b^* \in B$ and $x \in \bar{K}(b)$ be arbitrarily fixed. Then there are a point $y \in C(b)$ and a vector v with $\|v\| \leq \tau$ such that $x = y+v$. Since C, by Corollary 4.4. is Lipschitzian on B, there are a point $y^* \in C(b^*)$ and a constant number $\rho > 0$ (not depending on b and $b^*$) such that

$$\|y-y^*\| \leq \rho \|b-b^*\|.$$

By $C(b^*) \subset M(b^*)$ and (ii) there is a point $x^* \in G(b^*)$ with

$$\|x^*-y^*\| < \delta \ .$$

Now we have

$$\begin{aligned} \|x-x^*\| &= \|y+v-y^*+y^*-x^*\| \leq \\ &\leq \|y-y^*\| + \|v\| + \|y^*-x^*\| \leq \\ &\leq \rho \| b-b^*\| \quad + \tau + \delta \end{aligned}$$

and setting $\bar{\tau} = \tau + \delta$, the desired proposition follows. #

For the constraint sets G(b) from (15) we can discuss the question of a parametric description.

By Theorem 2. the constraint set G(b), for a fixed $b \in B$, has the representation

$$\left.\begin{aligned} &G(b) = K(b) + W, \\ &W = \{u \in R^n \ / \ u = \sum_{i=1}^{r} \beta_i v^i,\ \beta_i \text{ natural},\ i=1,\dots,r\}. \end{aligned}\right\} \qquad (16)$$

and, by Theorem 6., the set K(b) is a finite union of compact convex polyhedra (in the pure integer case K(b) consists of a finite number of integer points). If one now underlies that the polyhedra appearing in K(b) are given by the convex combinations of their vertex points, then (16) may be considered a parametric representation of the constraint set G(b). The validity of (16) is also guaranteed for a fixed $b \in B$ if one uses, in the construction of K(b) (due to the proof of Theorem 2.) and W, such a (mixed-) integer generator system $v^1, \dots, v^r$, for which the greatest common divisor of the integer components is equal to one and if the convex hull of the vertex points of the polyhedron $M(b) \cap H^{\perp}$ is taken instead of C(b) defined in (4.17), where H is the subspace of maximal dimension contained in the cone V.

Sometimes it is desirable to have a parametric representation of G(b) including a minimal number of unbounded parameters $\beta_i$ only. This minimal number coincides with the dimension of the polyhedral recession cone V, which one sees by applying a well-known result of CARATHÉODORY.

We consider the finite number of variations $k^1 = (k_1^1,\dots,k_d^1),\dots,k^t = (k_1^t,\dots,k_d^t)$ of the indices $1,\dots,r$, where $d = \dim V$ such that the corresponding vector systems $\{v^{k_1^j},\dots,v^{k_d^j}\}$ are linearly independent and such that $V = \bigcup_{j=1}^{t} V(k^j)$ holds, where

$$V(k^j) \underset{Df}{=} \{u \in R^n \,/\, u = \sum_{i=1}^{d} \beta_i v^{k_i^j},\ \beta_i \geq 0,\ i = 1,\dots,d\}.$$

Thereby one can choose the variations in such a way that the cones $V(k^j)$ have only common boundary points.

If one denotes by $K_{k^j}(b)$ the respective compact sets related to $K(b)$, then one has

$$G(b) = \bigcup_{j=1}^{t} \{K_{k^j}(b) + \{u \in R^n \,/\, u = \sum_{i=1}^{d} \beta_i v^{k_i^j}\ \ \beta_i \text{ natural},\ i=1,\dots,d\}\}.$$

Next, we will mention some properties of the domain B of G. Thereby we use the special description

$$G(b) = \{x \in R^n \,/\, x = (z,y),\ Az + Qy \leq b,\ z \in R^s,\ \text{integer}\} \tag{17}$$

where $D = (A,Q)$ is an $(m,n)$-matrix.

In Theorem 2.(iii) it was pointed out that the domain B of the multifunction G given by (3) is closed if the domain $\Lambda$ of the involved mapping M is closed and if the cone V satisfies (MIG). This result is also true for the multifunction G defined by (7) if the coefficients of the quasiconvex polynomial functions are rational numbers. However, if G is given by (17), then one gets more information about the structure of the domain B and, moreover, one may derive properties of the sets $B(b^o)$ defined in (8), on which G is continuous. These facts are listed in the following two remarks. The proofs can be found in BANK/BELOUSOV et al. (1986).

We shall use the following (known) definitions. Let $S \subset R^n$. Then S is called a <u>monoid</u> if $0 \in S$, and $u,v \in S$ implies that $u+v \in S$. The set S is said to be <u>star-shaped</u> if there exists a point $v \in S$ such that the line segment $s(u,v)$ between u and v belongs to S for every $u \in S$, $u \neq v$.

<u>Remark 1.</u>

Obviously, the domain B is a monoid. If the matrix (A,Q) is rational, then the domain B is closed and connected and may be represented as the union of a countable family of convex polyhedral cones which all are translations of the following fixed m-dimensional cone:

$$W \underset{Df}{=} \{g \in R^m \,/\, g \geq Qy \text{ for some } y \in R^{n-s}\}.$$

<u>Remark 2.</u>

If the matrix (A,Q) in (17) is rational, then the set $B(b^o)$ (8), $b^o \in B$, arbitrarily fixed, is star-shaped and neither open nor closed in general

There exist a finite set T of s-dimensional integer vectors z and a point $d^o \in R^m$ such that

$$B(b^o) = (d^o+W) \setminus \bigcup_{z \in T} (Az+W).$$

The monoid B possesses a complete decomposition into a countable number of sets $B(b^*)$ (8) $b^* \in B$.

We illustrate Remark 2. by the following figure and an example.

$B(b^o)$ 

Example 3.

$M(b) = \{x \in R^2 \ / \ 2x_1+x_2 \leq b_1, \ x_1+2x_2 \leq b_2\}$,
$b^o = (2,2)^T$.

(i) $G(b) = \{x \in M(b) \ / \ x_1, x_2 \text{ integer}\}$,
$B(b^o) = \{b \in R^2 \ / \ 2 \leq b_i < 3, \ i=1,2\}$.

(ii) $G(b) = \{x \in M(b) \ / \ x_1 \text{ integer}\}$,
$B(b^o) = R^2$.

Remark 3.

Using the parametric descriptions of the sets G(b) sketched in connection with the representation (16) one can also derive parametric representations for the domain B and the sets $B(b^o)$. Obviously, such a representation for the sets $B(b^o)$ will be of very complicated, not tractable form.

The next theorem will give a characterization of the vertex points of the convex hull of constraint sets having the form

$$G(b) = \{x \in R^n \ / \ Ax = b, \ x \geq 0, \ x_1,\ldots,x_s \text{ integer}\} \tag{18}$$

where A is an integer (m,n)-matrix. This latter assumption and the form of G(b) ensure that the convex hull is a convex polyhedron having a vertex point if it is non-empty.

Theorem 8.

Let G(b) be given by (18) and, let A have the rank m. Let $\Delta$ denote the maximum of the moduli of all m-order minors of A and $\delta$ their greatest common divisor. If $x^o$ is a vertex point of conv G(b), then there exists an index set $I = \{i_1,\ldots,i_m\} \subset \{1,\ldots,n\}$ such that the matrix $(A^{i_1},\ldots,A^{i_m})$ is regular and

$$x_i^o < \frac{\Delta}{\delta} \quad \forall i \notin I \tag{19}$$

holds where $A^{i_\alpha}$, $\alpha=1,\ldots,m$, represents the $i_\alpha$ -th column of A.

Proof.

First, one has that the vertex point $x^0$ belongs to $G(b)$. Obviously, the case $m = n$ is trivial. Therefore, we suppose that $m < n$. We consider the system of all index sets $J = \{j_1,\dots,j_m\} \subset \{1,\dots,n\}$ for which the matrix $(A^{j_1},\dots,A^{j_m})$ is regular, and we select from this system such an index set I for which the number of indices $i \notin I$ is maximal with respect to $x_i^0 < \frac{\Delta}{\delta}$. We show that the index set I chosen in such a way satisfies condition (19).

Let $i^* \notin I$ be such that $x^0_{i^*} \geq \frac{\Delta}{\delta}$. Without loss of generality we may assume that $I \cup \{i^*\} = \{1,\dots,m,m+1\}$. Now we consider the linear system

$$\sum_{i=1}^{m+1} A^i x_i = \sum_{i=1}^{m+1} A^i x_i^0, \ x_i \geq 0, \ i=1,\dots,m+1. \tag{20}$$

Since (20) has a solution, one concludes from linear programming that (20) has also a basic feasible solution. Without loss of generality we assume that $x_1,\dots x_m$ are the basic variables. By the choice of I we have $x^0_{m+1} \geq \frac{\Delta}{\delta}$. Using now CRAMER's rule one obtains the following representation of the basic variables

$$x_i = \frac{\alpha_i + \beta_i x_{m+1}}{\gamma}, \quad i = 1,\dots,m, \tag{21}$$

where

$$\alpha_i = \det(A^1,\dots,A^{i-1}, \sum_{i=1}^{m+1} A^i x_i^0, A^{i+1},\dots,A^m),$$

$$\beta_i = -\det(A^1,\dots,A^{i-1}, A^{m+1}, A^{i+1},\dots,A^m),$$

$$\gamma = \det(A^1,\dots,A^m).$$

The ratios $\frac{\beta_i}{\delta}$, for all $i=1,\dots,m$, and $\frac{\gamma}{\delta}$ are integers. Further, we have $|\gamma| \leq \Delta$, $|\beta_i| \leq \Delta$, $i=1,\dots,m$; and $\frac{\alpha_i}{\gamma} \geq 0$, $i=1,\dots,m$, holds because $(A^1,\dots,A^m)$ is a feasible basis in (20). Setting now

$$\bar{x}_{m+1} = x^0_{m+1} - \frac{|\gamma|}{\delta} \text{ and } \bar{\bar{x}}_{m+1} = x^0_{m+1} + \frac{|\gamma|}{\delta}$$

we obtain from (21) two points $\bar{x}, \bar{\bar{x}} \in R^{m+1}$.

Obviously, we have

$$\bar{x}_{m+1} \geq 0, \ \bar{\bar{x}}_{m+1} \geq 0$$

and, for $j \in \{1,\dots,m+1\}$,

$\bar{x}_j$ and $\bar{\bar{x}}_j$ are integers if $x_j^0$ is an integer number.

Further, $\bar{x}_i \geq 0$, $i=1,\dots,m$, since

$$x_i^0 = \frac{\alpha_i}{\gamma} + \frac{\beta_i}{\gamma} x^0_{m+1}, \ \bar{x}_{m+1} < x^0_{m+1}, \ \frac{\alpha_i}{\gamma} \geq 0.$$

We still check that

$$\bar{\bar{x}}_i \geq 0, \ i=1,\dots,m, \tag{22}$$

is satisfied. Let us assume that (22) is false. Then there is an index r such that

$$(A^1,\dots,A^{r-1}, A^{m+1}, A^{r+1},\dots,A^m)$$

is a feasible basis for (20) with

$$x_r = \frac{\alpha_r + \beta_r(x^o_{m+1} + \frac{|\gamma|}{\delta})}{\gamma} < 0.$$

Hence

$$0 \le x^o_r = \frac{\alpha_r + \beta_r\, x^o_{m+1}}{\gamma} < -\frac{|\gamma|}{\gamma} \cdot \frac{\beta_r}{\delta}$$

and it follows from $|\beta_r| \le \Delta$ that

$$x^o_r < \frac{\Delta}{\delta} .$$

But this contradicts the choice of the index set I. Therefore, (22) is valid.

Using now the shown properties of the points $\bar{x}$ and $\bar{\bar{x}}$ from $R^{m+1}$ we obtain, for the point $x^o$, the representation

$$x^o = 1/2\, x^1 + 1/2\, x^2$$

where the points

$$x^1 = (\bar{x}_1,\dots,\bar{x}_{m+1},x^o_{m+2},\dots,x^o_n),$$
$$x^2 = (\bar{\bar{x}}_1,\dots,\bar{\bar{x}}_{m+1},x^o_{m+2},\dots,x^o_n)$$

belong to G(b), which contradicts the assumption that $x^o$ is a vertex point of conv G(b). #

For the vertex points of the convex hull of G(b) (18) Theorem 8. exhibits a certain similarity to the well-known characterization of vertex points in linear programming. Of course, the bounds (19) are not very practicable because of the appearing constants $\Delta$ and $\delta$ . In order to illustrate the bounds we give the following example:

Example 4. For the set

$$G(11) = \{x \in R^2 \;/\; x_1+2x_2 = 11,\ x_1,x_2 \text{ natural}\}$$

we have $\Delta = 2$, $\delta = 1$ and

$$G(11) = \{\tbinom{1}{5}, \tbinom{3}{4}, \tbinom{5}{3}, \tbinom{7}{2}, \tbinom{9}{1}, \tbinom{11}{0}\}.$$

Besides the vertex points $\binom{1}{5}$ and $\binom{11}{0}$ of the convex hull of G(11) the point $\binom{9}{1}$ is the only one satisfying condition (19).

Corollary 4.

If the hypotheses of Theorem 8. are fulfilled and, in (18), all variables $x_i$ are **required** to be integers, then the following inclusion is valid:

$$\text{ext conv } G(b) \subset \{x \in R^n \;/\; x_{B_i} = B_i^{-1}(b-N_i v),\ x_{N_i} = v;\ v_j < \frac{\Delta}{\delta},\ v_j \text{ natural};\ j=1,\dots,n-m, i=1,\dots,q\}.$$

Thereby $B_i$, $i=1,\dots,q$, are all basic matrices of A, and $N_i$ denotes the matrix built up by the columns of A not belonging to $B_i$. $x_{B_i}$ and $x_{N_i}$ are subvectors of x associated with the columns of $B_i$ and $N_i$, respectively.

## 6. Stability Properties of Nonlinear Integer Optimization Problems

### 6.1. Introduction

This chapter is devoted to parametric (mixed-) integer optimization problems with a quadratic or a convex polynomial objective function and constraint sets G(b) of the form (5.3) or (5.7).
We shall start our investigation considering the following general parametric problem:

$$P(p,b) \quad \inf\{f(x) + p^T x \;/\; x \in G(b)\}\;, \; p \in R^n,\; b \in R^m, \tag{1}$$

where f(x) is a polynomial function on $R^n$ without a linear term, and the constraint sets G(b) are given by (5.3), i.e.

$G(b) = \{x \in M(b) \;/\; x_1,\dots,x_s \text{ integer}\}$
with
$M(b) = C(b) + V$; $C: \Lambda \longrightarrow 2^{R^n}$ a compact-valued u.s.c. multifunction defined on $\Lambda \subset R^m$ and V a polyhedral cone in $R^n$.

As already before, we denote by B the domain of the constraint-set mapping G.
Now, the aim is to characterize the continuity behaviour of the value function $\varphi$, the optimal set-function $\Psi$, both analogously defined as in section 3.1., and the $\varepsilon$-optimal set-function $\tilde{\Psi}$ given by

$$\tilde{\Psi}(\varepsilon,p,b) \underset{Df}{=} \{x \in G(b) \;/\; f(x) + p^T x \leq \varphi(p,b) + \varepsilon\} \tag{2}$$

where $\varepsilon \geqq 0$.

In section 6.2. we consider the fixed function $f: R^n \longrightarrow R$ in (1) to be a quadratic form and, in section 6.3. to be a convex polynomial function. Most continuity results are derived for the general class of constraint sets which are allowed in (1). But, in order to show the upper semicontinuity of the $\varepsilon$-optimal set-function $\tilde{\Psi}$ on certain subsets of its domain we underly the special constraint sets G(b) from (5.7) where the (CPC) condition is satisfied.

In our considerations the level-set-mapping of the problem P(p,b) plays an important role. This mapping is defined by the level sets

$$N(p,b,a) \underset{Df}{=} \{x \in R^n \;/\; x \in G(b),\; f(x)+p^T x \leq a\} \tag{3}$$

of the problem P(p,b). These sets define a multifunction, the level-set-mapping, $N: \mathcal{E}_N \longrightarrow 2^{R^n}$ with

$$\mathcal{E}_N \underset{Df}{=} \{(p,b,a) \in \mathcal{E} \times R \;/\; N(p,b,a) \neq \emptyset\}$$

where

$$\mathcal{E} \underset{Df}{=} \{(p,b) \in R^n \times B \;/\; \varphi(p\; b) > -\infty\}\;,$$

i.e. $\mathcal{E}$ is the set of all pairs (p,b) for which the constraint sets

$G(b) \neq \emptyset$ and the objective function is bounded from below. This set is called the domain of finiteness of $\varphi$.
As we shall see, the behaviour of the objective function on the fixed polyhedral cone V (appearing in the representation of the constraint G(b)) is of great importance in our developments. Therefore, we use the following definitions:

$$E(b) \underset{Df}{=} \{p \in R^n \,/\, \varphi(p,b) > -\infty\}, \; b \in B, \tag{4}$$

$$\varphi_V : F_0 \longrightarrow R \text{ with}$$
$$\varphi_V(p) \underset{Df}{=} \inf \{f(x)+p^Tx \,/\, x \in V\}, \; p \in R^n, \tag{5}$$
$$F_0 \underset{Df}{=} \{p \in R^n \,/\, \varphi_V(p) > -\infty\}.$$

## 6.2. Integer Problems with Quadratic Objective Function

We explore here the parametric problem P(p,b) (1) where the function f(x) has the form

$$f(x) = x^TDx$$

with a symmetric (n,n)-matrix D.
First, for the set

$$\tilde{V} \underset{Df}{=} \{x \in V \,/\, x^TDx = 0\} \tag{6}$$

we derive basic properties which will play an important role in our considerations. Thereby we make use of the following well-known necessary optimal condition (KUHN-TUCKER):

Lemma 1.
If $x^0$ is an optimal point of the quadratic program

$$\inf\{x^TDx + p^Tx \,/\, Ax \leq b\}$$

where D is a symmetric (n,n)-matrix and A an (m,n)-matrix, then:

(i) There exists a vector $u^0 \in R^m$ such that the pair $(x^0,u^0)$ satisfies the system

$$\begin{matrix} Ax \leq b, & -2Dx - A^Tu = p, \\ u \geq 0, & u^T(b-Ax) = 0. \end{matrix} \tag{7}$$

(ii) On the region (7) one has

$$x^TDx + p^Tx = \tfrac{1}{2}(p^Tx - b^Tu).$$

As a consequence of Lemma 1. we obtain a result of HANSEL (1980):

Lemma 2.
Let $x^TDx \geq 0$ for all $x \in V$. Then it holds:

(i) The set $\tilde{V}$ (6) is a union of a finite number of polyhedral cones.
(ii) There is a representation of $\tilde{V}$ as a finite union of polyhedral cones

which all satisfy the (IG) condition if V does so and if D is a rational matrix.

Proof. Without loss of generality we may underly the following representation for the polyhedral cone V:

$$V = \{x \in R^n \ / \ Ax \leqq 0\}$$

where one can suppose that the (m,n)-matrix A is rational if V satisfies the (IG) condition.

We consider the special quadratic problem

$$\inf\{x^TDx \ / \ Ax \leqq 0\}$$

where $x^TDx \geqq 0$ if $Ax \leqq 0$, by assumption.

Obviously, $\tilde{V}$ gives the optimal set of this quadratic problem. If now $x^0$ is an optimal point, then Lemma 1. implies the existence of a vector $u^0$ and an index-set $I(x^0)$ such that the pair $(x^0,u^0)$ satisfies the homogeneous linear system

$$\begin{aligned} & -2Dx - A^Tu = 0, \\ & u_i \geqq 0, \quad A_i^Tx = 0, \quad i \in I(x^0), \\ & u_i = 0, \quad A_i^Tx \geqq 0, \quad i \notin I(x^0), \end{aligned} \tag{8}$$

where $A_i$ denotes the i-th row of A.

Taking the proposition (ii) of Lemma 1. into account one has that $x \in \tilde{V}$ if the pair (x,u) satisfies the system (8). This system describes a polyhedral cone in $R^{n+m}$ whose projection onto $R^n$ also gives a polyhedral cone and, since there are only finitely many index-sets $I(x^0)$ possible, the proposition (i) is shown. The second statement is an immediate consequence of the rational data in (8). #

Agreement. We say "$\tilde{V}$ satisfies (MIG) (or (IG))" if $x^TDx \geqq 0$ for all $x \in V$ and if there are finitely many polyhedral cones satisfying (MIG) (or (IG)), whose union gives $\tilde{V}$.

Lemma 3.

The following propositions hold:

(i) If $x^TDx \geqq 0$ for all $x \in V$, then

$$F_o = \{p \in R^n \ / \ p^Tx \geqq 0 \ \forall x \in \tilde{V}\};$$

otherwise $F_o = \emptyset$.

(ii) $F_o$ is a polyhedral cone.

(iii) $\varphi_V$ is continuous on $F_o \neq \emptyset$.

Proof. (i) Well-known and due to EAVES (1971).

(ii) Let $F_o \neq \emptyset$. By Lemma 2., $\tilde{V}$ is a union of a finite number of polyhedral cones. Let $\{v^1,\ldots,v^h\}$ be a system of generators of these cones then, obviously,

$$F_o = \{p \in R^n \;/\; p^T v^k \geq 0,\; k = 1,\ldots,h\} .$$

(iii) The function $\varphi_V$ is concave on $F_o$ and by a well-known theorem (e.g. ROCKAFELLAR (1970)), it follows that $\varphi_V$ is lower semicontinuous on $F_o$. It remains to show that $\varphi_V$ is upper semicontinuous on $F_o$. Let $\{p^t\} \subset F_o$ be a sequence with $p^t \longrightarrow p^o$ and we choose an $\varepsilon > 0$. Then there exists a point $x^o \in V$ such that

$$x^{o^T} D x^o + p^{t^T} x^o < \varphi_V(p^o) + \varepsilon$$

for almost all t. Since $\varphi_V(p^t) \leq x^{o^T} D x^o + p^{t^T} x^o$, we have

$$\varphi_V(p^t) < \varphi_V(p^o) + \varepsilon \text{ for almost all } t,$$

and therefore, $\varphi_V(p^t) \leq \varphi_V(p^o)$ holds. #

<u>Lemma 4.</u>

If the cone V fulfils the (MIG) condition, then

$$F_o = F_s \underset{Df}{=} \{p \in R^n \;/\; \inf \{x^T D x + p^T x \;/\; x \in V_s\} > -\infty\}$$

where $V_s = \{x \in R^n \;/\; x_1,\ldots,x_s \text{ integer}\}$.

<u>Proof.</u> Obviously, one has only to show $F_o \supset F_s$. Furthermore, it is easy to see that we can consider the case

$$V = R^n_+ \text{ and } V_s = \{x \in R^n_+ \;/\; x_1,\ldots,x_s \text{ integer}\}$$

because the fulfilled (MIG) condition for the polyhedral cone V allows to apply a linear transformation of the variables using the mixed-integer generators of V.

We carry out the proof by contradiction.

Let us suppose that $p \in F_s \setminus F_o$. Then we know from Lemma 3.(i) that the following two cases are possible:

<u>Case 1.</u> There exists a point $x^* \in V$ such that $x^{*T} D x^* < 0$.

<u>Case 2.</u> There exists a point $x^* \in V$ such that $x^{*T} D x^* = 0$ and $p^T x^* < 0$.

In both cases we have the situation that

$$\inf\{x^T D x + p^T x \;/\; x \in S\} = -\infty$$

where

$$S \underset{Df}{=} \{x \in R^n \;/\; x = kx^*,\; k \geq 0\},$$

<u>Case 1.</u> Using the notations

$$x(k) \underset{Df}{=} [kx^*],$$
$$u(k) = kx^* - x(k)$$

we obtain, for $k \geq 0$,

$$x(k) \in R^n_+ \text{ and } x(k) \text{ integer}$$

and,

$$x(k)^T Dx(k) + p^T x(k) = k^2 x^{*'} Dx^* + k(-x^{*^T} Du(k) + p^T x^*) + \\ + u(k)^T Du(k) - p^T u(k).$$

Because of $\|u(k)\| \leq 1$ it follows that

$$x(k)^T Dx(k) + p^T x(k) \longrightarrow -\infty \quad \text{if } k \longrightarrow \infty.$$

Case 2. By Lemma 5.2., there is a sequence $\{x^t\} \subset R^n_+$ of integer points with $\|x^t\| \longrightarrow \infty$ and, $d(x^t,S) \longrightarrow 0$ if $t \longrightarrow \infty$. Therefore, we may write

$$x^t = z^t + u^t, \text{ for all } t,$$

where $z^t = k_t x^* \in S$, $k_t \longrightarrow \infty$ and $\|u^t\| \longrightarrow 0$.
(If S is contained in a face of $R^n_+$, then one applies Lemma 5.2. in the corresponding subspace.)
Then we have

$$x^{t^T} Dx^t + p^T x^t = z^{t^T} Dz^t + p^T x^t + 2z^{t^T} Du^t + u^{t^T} Du^t + p^T u^t = \\ = k_t(p^T x^* + 2x^{*^T} Du^t) + u^{t^T} Du^t + p^T u^t .$$

If t is sufficiently large, then the estimate

$$x^{t^T} Dx^t + p^T x^t \leq \frac{1}{2} k_t p^T x^* + u^{t^T} Du^t + p^T u^t$$

follows and, by $p^T x^* < 0$ and $k_t \longrightarrow \infty$, we obtain

$$x^{t^T} Dx^t + p^T x^t \longrightarrow -\infty \text{ if } k \longrightarrow \infty.$$

Thus, we have shown in both cases that $p \notin F_0$ leads to a contradiction. #

Our results obtained up to now enable us, in particular, to show that the value function is continuous if the constraint set is fixed.

Theorem 1.

If the condition (MIG) is satisfied for V, then, for every fixed $b \in B$:

(i) $\varphi(p,b) > -\infty \iff \inf\{x^T Dx + p^T x \;/\; x \in K(b) + V\} > -\infty$
$\iff \inf\{x^T Dx + p^T x \;/\; x \in \operatorname{conv} G(b)\} > -\infty;$

(ii) The set E(b) (4) is a convex polyhedron;

(iii) The value function $\varphi$ is continuous on $E(b) \neq \emptyset$.

Proof. (i) By Theorem 5.2 (i) we have conv G(b) = conv K(b) + V and, hence it suffices to show that

$$\inf\{x^T Dx + p^T x \;/\; x \in \operatorname{conv} K(b) + V\} = -\infty \Longrightarrow \varphi(p,b) = -\infty.$$

Let

$$\inf\{x^T Dx + p^T x \;/\; x \in \operatorname{conv} K(b) + V\} = -\infty \tag{9}$$

be valid. We shall show that there is a point $c^0 \in K(b)$ such that

$$\inf\{x^TDx + p^Tx \ / \ x \in c^o + V\} = -\infty. \tag{10}$$

For

$$\varphi^*(c) \underset{Df}{=} \inf\{v^TDv + (2Dc+p)^Tv \ / \ v \in V\}, \ c \in R^n,$$

we have

$$\inf\{x^TDx+p^Tx \ / \ x \in \text{conv } K(b)+V\} = \inf\{c^TDc+p^Tc+\varphi^*(c)/c \in \text{conv } K(b)\}.$$

By Lemma 3. (iii), $\varphi^*$ is continuous on the set $\{c \in R^n \ / \ \varphi^*(c) > -\infty\}$. Since conv K(b) is compact, it follows from (9) that there must be a $c^* \in$ conv K(b) such that $\varphi^*(c^*) = -\infty$.

If $v^TDv < 0$ for a $v \in V$, then every point from K(b) can figure as $c^o$. Therefore, let $v^TDv \geq 0$ for all $v \in V$. Then, by Lemma 3. (i) there exists a $\tilde{v} \in \tilde{V}$ with $(2Dc+p)^T\tilde{v} < 0$. Since $c^*$ may be expressed as a convex combination of a finite number of points belonging to K(b), there must also be a $c^o \in K(b)$ such that $(2Dc^o+p)^T\tilde{v} < 0$, which implies (10).

From (10) one concludes that also

$$\inf\{x^TDx + p^Tx \ / \ x \in c^o+V_s\} = -\infty$$

is valid, if one takes Lemma 3. (i) into account.

By $c^o + V_s \subset K(b) + V_s$ it now follows that $\varphi(p,b) = -\infty$.

(ii) If there is a $v \in V$ with $v^TDv < 0$, then, by (i), it follows that $E(b) = \emptyset$. Therefore, let $v^TDv \geq 0$, for all $v \in V$. Then, by Lemma 2., the set $\tilde{V}(6)$ is a union of finitely many polyhedral cones. Suppose that $\{\tilde{v}^1,\dots,\tilde{v}^h\}$ is a system of generators of all these cones. Then, taking (i) into account, one obtains

$$p \in E(b) \iff 2c^TD\tilde{v}^i + p^T\tilde{v}^i \geq 0 \ \forall c \in K(b), \quad i = 1,\dots,h,$$

which implies that

$$E(b) = \{p \in R^n \ / \ p^T\tilde{v}^i \geq \sup\{-2c^TD\tilde{v}^i \ / \ c \in K(b)\}, \ i = 1,\dots,h\}, \tag{11}$$

where E(b) must be a convex polyhedron.

The proposition (iii) can be shown similarly as (iii) in Lemma 3. #

For parametric quadratic optimization problems without integer requirements on the variables one can point out the description of the polyhedron E(b) (11) more precisely.

<u>Remark 1.</u>

Let $G(b) \equiv M(b)$ (i.e. s = 0) and let M(b) be given by (4.5), where (CPC) is satisfied. Then the multifunction $E: B \longrightarrow 2^{R^n}$ defined by

$$E(b) \underset{Df}{=} \{p \in R^n \ / \ p^T\tilde{v}^i \geq \eta_i(b), \ i = 1,\dots,h\}$$

is continuous on B ( $= \Lambda$ ). Here the functions

$$\eta_i(b) \underset{Df}{=} \sup\{-2c^TD\tilde{v}^i \ / \ c \in C(b)\}, \ i = 1,\dots,h,$$

where C(b) is taken from Theorem 4.3., are continuous because of the continuity of C.

Remark 2.

If one additionally requires in Remark 1. that the involved polynomial functions are only linear, i.e.

$$G(b) = \{x \in R^n \,/\, Ax \leq b\}$$

where A is an (m,n)-matrix, then the domain of finiteness $\mathcal{E}$ of $\varphi$ is a union of finitely many polyhedral cones. In order to see this we recall from parametric linear programming the two facts that B is a convex polyhedral cone which may be divided into a finite number of convex polyhedral cones and, on which each of the functions $\eta_i(b)$ (from Remark 1.) is linear (cf. NOŽIČKA et al. (1974)). Consequently, each of the sets

$$W_i \underset{Df}{=} \{(p,b) \in R^n \times B \,/\, p^T\tilde{v}^i \geq \eta_i(b)\}\ ,\ i=1,\dots,h,$$

is a union of finitely many convex polyhedral cones, which implies that the same also holds for

$$\mathcal{E} = \bigcap_{i=1}^{h} W_i .$$

The following lemma is a powerful tool for treating the general quadratic problem P(p,b) with constraint sets G(b) given by (5.3). It should be noted that this lemma was established first in HANSEL (1980) for the case of linear (mixed-)integer constraints.

Lemma 5.

Let $\mathcal{E} \neq \emptyset$ and let $\tilde{V}$ satisfy (MIG) and furthermore let $(p^t,b^t,a_t) \in \mathcal{E}_N$, t = 1,2,..., be a sequence with

$$\lim(p^t,b^t,a_t) = (p^o,b^o,a_o)$$

where $b^o \in B$. Then, for every sequence $\{x^t\} \subset R^n$ having the properties $x^t \in N(p^t,b^t,a_t)$ and $\|x^t\| \longrightarrow \infty$, there is a vector $v \in V_s$, $v \neq 0$, such that $v^TDv = 0$. Further, there is a subsequence $\{y^r\}$ of the sequence $\{y^t\}$ defined by $y^t \underset{Df}{=} x^t - v$ such that the following three statements hold:

(i) $y^r \in N(p^r,b^r,a_r)$;

(ii) $\|y^r\| < \|x^r\|$;

(iii) For every natural number k there is an index r(k) with

$$x^r - kv \in N(p^r,b^r,a_r)\quad \forall r > r(k).$$

Finally, if $\tilde{V}$ satisfies (IG), then v can be considered to be an integer vector.

Proof. We can suppose that the convex polyhedral cone V is described in the form

$$V = \{v \in R^n \,/\, {d^i}^T v \leq 0,\ i = 1,\dots,q\},$$

and, that the elements of $\{x^t\}$ may be represented by

$$x^t = z^t + \alpha_t u^t \tag{12}$$

where $u^t \in V$, $\|u^t\| = 1$, $z^t \in C(b^t)$, $\alpha_t \longrightarrow \infty$.

Without loss of generality* we can assume that $u^t \longrightarrow u^o$ where $u^o \in V$ and $\|u^o\| = 1$. Because of the upper semicontinuity of C and since $C(b^o) \neq \emptyset$ is a compact set, the sequence $\{z^t\}$ must be bounded. The inequality

$$x^{t^T} Dx^t + p^{t^T} x^t = z^{t^T} Dz^t + p^{t^T} z^t + \alpha_t (2Dz^t + p^t)^T u^t + \alpha_t^2 u^{t^T} Du^t \leq a_t$$

yields $u^{o^T} Du^o = 0$ if one takes into account that $u^{t^T} Du^t \geq 0$.
Using the index-set $I \underset{Df}{=} \{i \in \{1,\dots,q\} / \ d^{i^T} u^o = 0\}$ we obtain

$$u^o \in \tilde{V},\ \|u^o\| = 1,\ d^{i^T} u^o = 0 \quad \forall i \in I.$$

Since $\tilde{V}$ satisfies (MIG), and because of $d^{i^T} v \leq 0$, for all $i \in I$ and all $v \in V$, there must be a vector $v \in R^n$ whose components $v_1, \dots, v_s$ are integers such that

$$v \in \tilde{V},\ \|v\| \geq 1,\ u^{o^T} v \geq \frac{1}{2},\ d^{i^T} v = 0 \ \forall i \in I.$$

If $\tilde{V}$ satisfies (IG), then v can be considered to be an all integer vector.
Setting now $y^t \underset{Df}{=} x^t - v$ one has

$$\|y^t\|^2 = \|x^t\|^2 + \|v\|^2 - 2x^{t^T} v = \|x^t\|^2 + \|v\|^2 - 2z^{t^T} v - 2\alpha_t u^{t^T} v$$

and, taking into account that $\alpha_t \to \infty$, $v^T u^o \geq \frac{1}{2}$ and $u^t \longrightarrow u^o$ hold, we obtain without loss of generality* that $\|y^t\| < \|x^t\|$, which proves the assertion (ii).
For a fixed t we now consider the points $x = x^t - \beta v$, $\beta \geq 0$.
Substituting (12) one has $x = z^t + \alpha_t u^t - \beta v$, $\beta \geq 0$, and by $z^t \in C(b^t)$ it follows that

$$z^t + \alpha_t u^t - \beta v \in C(b^t) + V$$

if it is ensured that $\alpha_t u^t - \beta v$ belongs to V. In order to discuss the latter question we distinguish between two cases:

1) If $d^{i^T} v = 0$, then we have

$$d^{i^T}(\alpha_t u^t - \beta v) = \alpha_t\, d^{i^T} u^t \leq 0 \quad \forall \beta \geq 0.$$

2) If $d^{i^T} v < 0$, then the choice of v implies that $d^{i^T} u^o < 0$ holds. By $u^t \longrightarrow u^o$ we have without loss of generality*
$d^{i^T} u^t \leq \frac{1}{2} d^{i^T} u^o < 0$, and by

$$d^{i^T}(\alpha_t u^t - \beta v) = \alpha_t d^{i^T} u^t - \beta d^{i^T} v \leq \frac{1}{2}\alpha_t d^{i^T} u^o - \beta d^{i^T} v,$$

one finally obtains

$$d^{i^T}(\alpha_t u^t - \beta v) \leq 0 \quad \forall \beta \in \left[0, \frac{1}{2}\alpha_t \frac{d^{i^T} u^o}{d^{i^T} v}\right]$$

For every natural number k it holds that the components $x_j^t - kv_j$, $j = 1,\dots,s$, are integer numbers. Now, by $\alpha_t \longrightarrow \infty$ the discussion of both cases above implies that without loss of generality* $x^t - v \in G(b^t)$ and, for every natural k, there is an index r(k) such that

$$x^t - kv \in G(b^t) \quad \forall t > r(k).$$

The proof of (i) and (iii) will be completed by estimating as follows:

$$(x^t - kv)^T D(x^t - kv) + p^{t^T}(x^t - kv) =$$

$$= x^{t^T} Dx^t + p^{t^T} x^t - k(2Dx^t + p^t)^T v \leq$$

$$\leq x^{t^T} Dx^t + p^{t^T} x^t \leq a_t \quad \forall k \geq 0,$$

where we took into account that $(p^t, b^t) \in \mathcal{E}$ implies the inequality

$$(2Dx^t + p^t)^T v \geq 0. \quad \#$$

The next theorem will be devoted to the existence of optimal points for the general quadratic problem P(p,b) with constraint sets of the type (5.3).

Theorem 2.
If the domain of finiteness $\mathcal{E}$ is non-empty and $\tilde{V}$ satisfies (MIG), then the equivalence

$$P(p,b) \text{ has an optimal solution } \Longleftrightarrow (p,b) \in \mathcal{E}$$

is true.

Proof. Obviously, one has only to show that $(p,b) \in \mathcal{E}$ implies the existence of an optimal point of P(p,b). Let $(p,b) \in \mathcal{E}$ be a fixed parameter pair. Setting

$$a_o \underset{Df}{=} \varphi(p,b) \quad \text{and} \quad a_t \underset{Df}{=} a_o + \frac{1}{t}$$

we can choose points

$$x^t \in N(p,b,a_t), \; t = 1,2,\dots,$$

of minimal norm. By Lemma 5., the sequence $\{x^t\}$ must have an accumulation point which belongs to $N(p,b,a_o)$. #

Now we present two conditions under which the value function $\varphi$ is l.s.c. where the condition (a), by Theorem 2., even ensures that the problem P(p,b) is solvable.

Theorem 3.

The value function $\varphi$ is l.s.c. on $\mathcal{E} \neq \emptyset$ if one of the two conditions

(a) $\tilde{V}$ satisfies (MIG),

(b) V satisfies (MIG)

is fulfilled.

Proof. (By contradiction.)

We assume that $\varphi$ is not l.s.c. at $(p^o,b^o) \in \mathcal{E}$. Then there exists a sequence $(p^t,b^t) \in \mathcal{E}$, $t = 1,2,\ldots$, $(p^t,b^t) \longrightarrow (p^o,b^o)$ such that

$$\varphi(p^t,b^t) < \varphi(p^o,b^o) - \varepsilon \tag{13}$$

for some $\varepsilon > 0$.

Let the condition (a) be fulfilled. By Theorem 2., one has

$$N(p^t,b^t,\ \varphi(p^t,b^t)) \neq \emptyset,\ t = 1,2,\ldots\ .$$

Now we consider a sequence $\{x^t\}$ where $x^t$ is an element of minimal norm belonging to the level set $N(p^t,b^t,\ \varphi(p^t,b^t))$.

If this sequence has an accumulation point $x^o$, then we obtain from (13) the contradiction

$$\varphi(p^o,b^o) \leq x^{o^T}Dx^o + p^{o^T}x^o \leq \varphi(p^o,b^o) - \varepsilon$$

where $x^o \in G(b^o)$ is taken into account.

On the other hand, if $\|x^t\| \longrightarrow \infty$, then by Lemma 5., there exists a vector $v \in V_s$ with $v^TDv = 0$ such that, without loss of generality*,

$$x^t - v \in G(b^t),\ \|x^t - v\| < \|x^t\|. \tag{14}$$

Since $x^t + v \in G(b^t)$ and $x^t$ is an optimal solution of $P(p^t,b^t)$, we can estimate

$$\begin{aligned}\varphi(p^t,b^t) &\leq (x^t+v)^T D(x^t+v) + p^{t^T}(x^t+v) = \\ &= \varphi(p^t,b^t) + (2Dx^t+p^t)^T v.\end{aligned}$$

But, this is only possible for the case $(2Dx^t+p^t)^Tv \geq 0$. Consequently,

$$(x^t-v)^TD(x^t-v) + p^{t^T}(x^t-v) = \varphi(p^t,b^t) - (2Dx^t+p^t)^Tv \leq \varphi(p^t,b^t),$$

from which one obtains a contradiction to the choice of $x^t$ if one takes (14) into account. Now, let the condition (b) be satisfied.

From Theorem 5.2. we know that G(b) has a representation

$$G(b) = K(b) + V_s,\ b \in \mathcal{B},$$

where K is a compact-valued u.s.c. multifunction on $\mathcal{B}$. We define

$$\tilde{\varphi}(c,p) \underset{Df}{=} \inf\{v^TDv + (2Dc+p)^Tv \ /\ v \in V_s\}.$$

By (13) there are points $c^t \in K(b^t)$, $t = 1,2,\ldots$, having the property

$$c^{t^T}Dc^t + p^{t^T}c^t + \tilde{\varphi}(c^t,p^t) < \varphi(p^o,b^o) - \frac{\varepsilon}{2}. \tag{15}$$

The sequence $\{c^t\}$ has an accumulation point $c^o \in K(b^o)$ and $c^t \longrightarrow c^o$ holds without loss of generality*. Theorem 1. (iii) implies $\lim \tilde{\varphi}(c^t,p^t) = \tilde{\varphi}(c^o,b^o)$, which yields, using (15), the inequality

$$\varphi(p^o,b^o) \leq \varphi(p^o,b^o) - \frac{\varepsilon}{2}.$$

But this is impossible and, consequently, $\varphi$ must be l.s.c. #

Next, we focus our interest on the level-set mapping $N\colon \mathcal{E}_N \longrightarrow 2^{R^n}$ itself. However, to be able to establish its upper semicontinuity in general one has to impose the stronger hypotheses that both the cone V and the set $\tilde{V}$ have to satisfy the (IG) condition.

Theorem 4.

Let D be a rational matrix and let $\tilde{V}$ satisfy (IG). If $(p^t,b^t,a_t) \in \mathcal{E}_N$, $t = 1,2,\ldots$, is a sequence with $(p^t,b^t,a_t) \to (p^o,b^o,a_o)$ where $(p^o,b^o) \in \mathcal{E}$, then the following statement holds for the level sets:

$$\forall \varepsilon > 0 \;\; \exists t(\varepsilon)\colon N(p^t,b^t,a_t) \subset U_\varepsilon N(p^o,b^o,a_o) \; \forall t > t(\varepsilon).$$

Proof. We suppose that the theorem is false. Then there is, among all counterexamples, one for which the convex hull conv $G(b^o)$ has minimal dimension. Let us consider such a counterexample.
Without loss of generality* we can assume that

$$A_t \underset{Df}{=} N(p^t,b^t,a_t) \setminus U_\varepsilon N(p^o,b^o,a_o) \neq \emptyset,$$

for $t = 1,2,\ldots$, and some $\varepsilon > 0$. We choose an element $x^t$ of minimal norm in $A_t$, for all t. We may suppose that $\|x^t\| \longrightarrow \infty$ because, if the sequence $\{x^t\}$ had an accumulation point $x^o$, it would belong to $N(p^o,b^o,a_o)$, implying that $A_t$ would be empty for almost all t. By Lemma 5., there exists an integer vector $v \in V_s$, $v \neq 0$, such that $v^TDv = 0$ and there is a subsequence $\{y^r\}$, $y^r = x^r - v$, corresponding to $\{x^r\}$ and possessing the properties (i), (ii) and (iii) of Lemma 5. Because of (i) and (ii) (of Lemma 5.) and on account of the minimality of $x^r$,

$$y^r \in U_\varepsilon N(p^o,b^o,a_o) \;\; \forall r$$

must be valid. Then there are points $z^r \in N(p^o,b^o,a_o)$ fulfilling $\|y^r - z^r\| < \varepsilon$ and, by $z^r + v \in G(b^o)$ and $\|x^r - (z^r+v)\| < \varepsilon$, we obtain the estimation

$$\begin{aligned} a_o &< (z^r+v)^TD(z^r+v) + p^{o^T}(z^r+v) = \\ &= z^{r^T}Dz^r + p^{o^T}z^r + (2Dz^r+p^o)^Tv \leq a_o + (2Dz^r+p^o)^Tv \end{aligned}$$

Hence

$$(2Dz^r+p^o)^Tv > 0.$$

Therefore

$$\dim \operatorname{conv} G(b^o) > \dim \operatorname{conv} \{x \in G(b^o) \;/\; 2x^T D v = -p^{o^T} v\} \tag{16}$$

is true.
From $x^r + kv \in G(b^r)$, for all natural numbers k, $(p^r, b^r) \in \mathcal{E}$ and

$$(x^r+kv)^T D(x^r+kv) + p^{r^T}(x^r+kv) =$$
$$= x^{r^T} D x^r + p^{r^T} x^r + k(2Dx^r+p^r)^T v$$

one obtains

$$(2Dx^r+p^r)^T v \geq 0. \tag{17}$$

Now we demonstrate that indeed

$$\lim (2Dx^r+p^r)^T v = 0. \tag{18}$$

Otherwise it would be $(2Dx^r+p^r)^T v \geq \alpha > 0$, for infinitely many indices r, and, using Lemma 5. (iii), we could conclude, for an arbitrary natural number k, that

$$(x^r-kv)^T D(x^r-kv) + p^{r^T}(x^r-kv) \leq$$
$$\leq x^{r^T} D x^r + p^{r^T} x^r - k\alpha \leq a_r - k\alpha$$

and $x^r - kv \in G(b^r)$ if $r > r(k)$. Then it would follow, for k sufficiently large and $a_r \longrightarrow a_o$, that $\underline{\lim}\, \varphi(p^r, b^r) = -\infty$, which would be impossible because of $(p^o, b^o) \in \mathcal{E}$ and the fact that $\varphi$ is lower semicontinuous on $\mathcal{E}$
Therefore (18) holds, which together with $p^r \longrightarrow p^o$ implies that

$$\lim 2x^{r^T} D v = -p^{o^T} v. \tag{19}$$

Now let us define

$$\hat{G}(b,\beta) \underset{Df}{=} \{x \in G(b) \;/\; 2x^T D v = \beta\},$$
$$\hat{B} \underset{Df}{=} \{(b,\beta) \in B \times R \;/\; \hat{G}(b,\beta) \neq \emptyset\},$$
$$\hat{V} \underset{Df}{=} \{u \in V \;/\; 2\, u^T D v = 0\}.$$

By Theorem 2.5., there exists a compact-valued u.s.c. multifunction $\hat{C} : (B \times R) \longrightarrow 2^{R^n}$ such that

$$\hat{G}(b,\beta) = \{x \in R^n \;/\; x \in \hat{C}(b,\beta) + \hat{V},\ x_1, \ldots, x_2 \text{ integer}\}$$

holds, for all $(b,\beta) \in \hat{B}$. For the multifunction $\hat{G}$ we declare the sets $\hat{\mathcal{E}}$ and $\hat{\mathcal{E}}_{\hat{N}}$ and the mapping $\hat{N}$ in an analogous way as in the case of the multifunction G. Because of the rationality of D and as v is an all integer vector, the set $\{u \in \hat{V} \;/\; u^T D u = 0\}$ satisfies, by Lemma 6.2., the condition (IG). If we set now $\beta_o \underset{Df}{=} -p^{o^T} v$ and $\beta_r \underset{Df}{=} 2v^T D x^r$, then it follows from (19) that

$$\lim(p^r, (b^r, \beta_r), a_r) = (p^o, (b^o, \beta_o), a_o).$$

Furthermore,

$$(p^r,(b^r,\beta_r),a_r)\in\hat{\mathcal{E}} \quad \text{and}$$

$$x^r\in\hat{N}(p^r,(b^r,\beta_r),a_r)\setminus U_{\varepsilon}\hat{N}(p^o,\ (b^o,\beta_o),a_o).$$

This says that we have constructed a further counterexample for which the statement of this theorem is false. However, by (16) it holds

$$\dim\operatorname{conv} G(b^o,\beta_o) < \dim\operatorname{conv} G(b^o),$$

which contradicts the choice of the original counterexample. #

Corollary 1.

If the assumptions of Theorem 4. are fulfilled, then $N(p^o,b^o,a_o)\neq\emptyset$.

Proof.

We choose $x^t$ such that

$$\|x^t\| = \min\{\|x\| \ / \ x\in N(p^t,b^t,a_t)\}$$

for $t = 1,2,\ldots$ . By Theorem 4., the sequence $\{x^t\}$ has an accumulation point $x^o$ because otherwise one would obtain a contradiction to the choice of $x^t$. Then one immediately sees that $x^o\in N(p^o,b^o,a_o)$. #

In Theorem 5. we show that the $\varepsilon$-optimal set-function $\tilde{\Psi}$ (2) possesses the property to be u.s.c. on a certain subset of $\mathcal{E}^*\times R_+$ if besides the hypotheses of Theorem 4. one additionally requires that the value function $\varphi$ is continuous on the subset $\mathcal{E}^*$ of its domain of finiteness $\mathcal{E}$ . On account of Theorem 3., it is in fact the requirement that $\varphi$ is u.s.c. on $\mathcal{E}^*$.

Theorem 5.

Let D be a rational matrix and let $\tilde{V}$ satisfy (IG). If $\mathcal{E}^*$ is a subset of $\mathcal{E}$ on which $\varphi$ is continuous, then the $\varepsilon$-optimal set-function $\tilde{\Psi}$ is u.s.c. on $\mathcal{E}^*\times R_+$.

Proof.

Let $(p^t,b^t,\varepsilon_t)\in\mathcal{E}^*\times R_+$, $t = 1,2,\ldots,$ be a sequence which converges to $(p^o,b^o,\varepsilon_o)$. Then, for the vectors $(p^t,b_t,a_t)$ with $a_t \underset{Df}{=} \varphi(p^t,b^t)+\varepsilon_t$ we have

$(p^t,b^t,a_t)\longrightarrow(p^o,b^o,a_o)$ and $(p^t,b^t,a_t)\in\mathcal{E}_N$, for $t = 0,1,2,\ldots$ .

Since

$$\tilde{\Psi}(p^t,b^t,\varepsilon_t) = N(p^t,b^t,a_t),\ t = 0,1,2,\ldots,$$

one obtains the proposition by Theorem 4. #

Now we consider the quadratic parametric (mixed-) integer optimization problem

$$P(p,b) \quad \inf\{x^TDx + p^Tx \;/\; x \in G(b)\} \;,\; b \in \mathcal{B},$$

where the constraint sets $G(b)$ are given in the form

$$G(b) = \{x \in R^n \;/\; p_i(x) \leq b_i,\; i=1,\dots,m;\; x_1,\dots,x_s \text{ integer}\} \tag{20}$$

with quasiconvex polynomial functions $p_i(x)$, $i = 1,\dots,m$.

This kind of constraint sets we have investigated in section 5.2. For the case that the involved sets

$$M(b) = \{x \in R^n \;/\; p_i(x) \leq b_i,\; i = 1,\dots,m\}$$

satisfy the condition (CPC) and their common recession cone $V = 0^+M(b)$ possesses the (MIG) property we could identify subsets $B(b^o)$ (5.8) of the domain of the multifunction G on which G is continuous. Employing this sets $B(b^o)$ we can expose subsets $\mathcal{E}^* \subset \mathcal{E}$ on which the continuity of the value function $\varphi$, assumed in Theorem 5., is ensured.

Theorem 6.

Let the constraint sets $G(b)$ be given by (20) and let (CPC) be satisfied for M. Further, let $b^o \in \mathcal{B}$ be a fixed parameter point and $\mathcal{E}^*$ given by

$$\mathcal{E}^* \underset{Df}{=} \{(p,b) \in \mathcal{E} \;/\; b \in B(b^o)\}$$

where $B(b^o)$ is defined in (5.8).

Then

(i) The value function $\varphi$ is continuous on $\mathcal{E}^*$ if the recession cone $V = 0^+M(b)$ satisfies (MIG).

(ii) The $\varepsilon$ -optimal set-function $\tilde{\Psi}$ is u.s.c. on $\mathcal{E}^* \times R_+$ if the matrix D is rational and the recession cone V satisfies (IG).

Proof.

We have only to show (i) and here, because of Theorem 3., it suffices to verify that $\varphi$ is upper semicontinuous on $\mathcal{E}^*$. Let $(p^o,b^o) \in \mathcal{E}^*$ be arbitrarily fixed, $\varepsilon > 0$ and $x^o \in G(b^o)$ such that

$$x^{o^T}Dx^o + p^{o^T}x^o < \varphi(p^o,b^o) + \varepsilon \;.$$

We assume that $(p^t,b^t) \longrightarrow (p^o,b^o)$, By Theorem 5.4. (ii), the constraint-set mapping G is l.s.c. at $b^o$ and therefore, there are points $x^t \in G(b^t)$, for t sufficiently large, such that $x^t \longrightarrow x^o$. Hence, one has

$$\overline{\lim}\; \varphi(p^t,b^t) \leq \overline{\lim}(x^{t^T}Dx^t + p^{t^T}x^t) =$$

$$= x^{o^T}Dx^o + p^{o^T}x^o < \varphi(p^o,b^o) + \varepsilon \;,$$

from which it follows that $\varphi$ is upper semicontinuous at $(p^o,b^o)$. #

## 6.3. Integer Problems with a Convex Polynomial Objective Function

In this section we treat parametric (mixed-) integer problems P(p,b) (1) where the objective function is a convex polynomial function with a variable linear form, i.e. the function f(x) in (1) is a convex polynomial function without a linear term. The way we proceed is similar to that in section 6.2.

The polyhedral cone

$$\tilde{V} \underset{Df}{=} K(f) \cap V \tag{21}$$

where $K(f) \underset{Df}{=} \{x \in R^n \ / \ \sup\{f(tx) \ / \ t \geq 0\} < \infty\}$

will play the same role as the set $\{x \in V \ / \ x^T Dx = 0\}$ in the previous section. By section 4.1., we know:

1. $K(f) = 0^+ \{x \in R^n \ / \ f(x) \leq 0\} = R(f)$ is a linear subspace (this holds since f has no linear term) and therefore,

   $$f(x+kv) = f(x) \quad \forall x \in R^n \, \forall k \in R \, \forall v \in K(f). \tag{22}$$
2. The linear subspace K(f) may be represented as the solution set of a system of homogeneous linear equations having only integer coefficients if all the coefficients of the polynomial function f are rational numbers.

The following lemma yields a tool which may be used to characterize the domain of finiteness E(b) and the value function $\varphi$ if the parameter $b \in B$ of the constraints is fixed. We still need the definition

$$\varphi_o(p,b) \underset{Df}{=} \inf\{f(x) + p^T x \ / \ x \in C(b) + V\} .$$

Lemma 6.

If f(x) is a convex polynomial function without a linear term and $\varphi(p,b) = -\infty$, then there exists a vector $v \in \tilde{V}$ (21) such that $p^T v < 0$.

Proof.

If the proposition is false, then there is a counterexample for which the cone V has minimal dimension. Then we have for such a counterexample that

$$p^T v \geq 0 \quad \forall v \in \tilde{V}. \tag{23}$$

We choose points $x^t \in C(b) + V$, $t = 1,2,\ldots$, with $f(x^t)+p^T x^t \longrightarrow -\infty$ which may be represented in the form

$$x^t = c^t + \alpha_t u^t \text{ with } c^t \in C(b),\ u^t \in V,\ \|u^t\| = 1,\ \alpha_t \longrightarrow \infty,$$

we suppose without loss of generality* that

$u^t \longrightarrow u^o$, $\|u^o\| = 1$, $u^o \in V$,

and $x^t \in A \underset{Df}{=} \{x \in R^n \,/\, f(x) + p^T x \leq 0\}$ for all t.

A well-known fact from convex analysis (cf. ROCKAFELLAR (1970)) says that $u^o \in 0^+A$. On account of Lemma 4.5. and Lemma 4.4. one concludes $u^o \in K(f)$ and $p^T u^o \leq 0$. Because of (23) this gives

$$u^o \in V, \ \|u^o\| = 1, \ \ p^T u^o = 0. \tag{24}$$

Now we consider the problem

$$\inf\{f(x) + p^T x \,/\, x \in C(b) + \hat{V}\} \tag{25}$$

where $\hat{V} \underset{Df}{=} \{u \in V \,/\, u^{o^T} u = 0\}$. Taking $\varphi_o(p,b) = -\infty$ into account we obtain from (24) and (25) that

$$\inf\{f(x) + p^T x \,/\, x \in C(b) + \hat{V}\} = -\infty$$

must hold. Since $u^o \notin \hat{V}$, it follows that $\dim \hat{V} < \dim V$. Thus, the counterexample (25) yields a contradiction to the choice of the original counterexample. #

Corollary 2.

When $b \in \Lambda$, then we have

$$\varphi_o(p,b) = -\infty \Longleftrightarrow \varphi_V(p) = -\infty$$

where $\varphi_V(p)$ is defined by (5).

Proof. One concludes this proposition from (22) and Lemma 6. #

Corollary 3.

When f(x) is a convex polynomial function without a linear term, $b \in B$, and the cone $\tilde{V}$ satisfies the condition (MIG), then the equivalence

$$\varphi_o(p,b) = -\infty \Longleftrightarrow \varphi(p,b) = -\infty$$

holds.

Proof. Obviously, it suffices to show the implication " $\Longrightarrow$". Let $x \in G(b)$ be arbitrarily chosen. Since $\tilde{V}$ satisfies (MIG) and because of Lemma 6. there is a generator v of $\tilde{V}$ with $p^T v < 0$, and $v_1, \dots, v_s$ are integer numbers.

For each natural number k we have $x + kv \in G(b)$ and $f(x+kv) + p^T(x+kv) = f(x) + p^T x + kp^T v$, from which $\varphi(p,b) = -\infty$ follows. #

Theorem 7.

Let f(x) be a convex polynomial function without a linear term and let $\tilde{V}$ satisfy (MIG). Then:

(i) The domain of finiteness $E(b)$ is a polyhedral cone not depending on $b \in B$, and it has the representation

$$E(b) = \{p \in R^n \ / \ p^T v \geq 0, \ v \in \tilde{V}\}.$$

(ii) The value function $\varphi(.,b)$ is continuous on $E(b)$.

Proof. One obtains (i) immediately from Lemma 6. and the Corollaries 2. and 3. To show the continuity of $\varphi(.,b)$ on the polyhedron $E(b)$ one proceeds as it was done in the proof of Lemma 3. for the value function $\varphi_V(p)$. #

The following Lemma will be a basic tool for the investigation of the general problem $P(p,b)$ (1) with a convex polynomial objective function; it is of the same importance as Lemma 5. in the previous section.

Lemma 7.

Let $f(x)$ be a convex polynomial function without a linear term and let the cone $\tilde{V}$ satisfy the condition (MIG). Further, let $(p^t,b^t,a_t) \in \mathcal{E}_N$, $t = 1,2,\ldots$, be a sequence with

$$\lim(p^t,b^t,a_t) = (p^o,b^o,a_o) \text{ and } b^o \in B.$$

Then, for every sequence $\{x^t\} \subset R^n$ having the properties $x^t \in N(p^t,b^t,a_t)$ and $\|x^t\| \longrightarrow \infty$, there is a vector $v \in \tilde{V}$, $v \neq 0$, $v_1,\ldots,v_s$ integer, such that the sequence $\{y^t\}$ defined by $y^t \underset{Df}{=} x^t - v$ contains a subsequence $\{y^r\}$ for which the following assertions hold:

(i) $y^r \in N(p^r,b^r,a_r)$;

(ii) $\|y^r\| < \|x^r\|$ ;

(iii) For every natural number $k$ there is an index $r(k)$ such that

$$x^r - kv \in N(p^r,b^r,a_r) \ \forall \ r > r(k).$$

Proof. For the cone $V$ we can assume that it is decribed by

$$V = \{v \in R^n \ / \ d^{i^T} v \leq 0, \ i = 1,\ldots,q\}.$$

The points $x^t$ may be written in the form

$$x^t = z^t + \alpha_t u^t \quad \text{with } u^t \in V, \ \|u^t\| = 1, \ z^t \in C(b^t), \alpha_t \longrightarrow \infty.$$

Further, without loss of generality* we can assume that $u^t \longrightarrow u^o$ where $\|u^o\| = 1$ and $u^o \in V$. Since the multifunction $C$ is upper semicontinuous and the non-empty set $C(b^o)$ is compact, the sequence $\{z^t\}$ must be bounded. The set

$$A \underset{Df}{=} \{(x,x_{n+1}) \in R^{n+1} \ / \ f(x) + x_{n+1} \leq 0\}$$

is closed and convex and we have

$$x^t \in N(p^t, b^t, a_t) \Longrightarrow (x^t, p^{t^T} x^t - a_t) \in A,$$

$$\frac{1}{\alpha_t}(x^t, p^{t^T} x^t - a_t) \longrightarrow (u^o, p^{o^T} u^o).$$

As in the proof of Lemma 6. we can conclude that $(u^o, p^{o^T} u^o) \in 0^+A$, from which

$$f(ku^o) + (k \cdot u^o)^T p^o \leq 0 \quad \forall k \geq 0$$

follows and therefore,

$$f(ku^o) = 0 \; \forall k \geq 0, \quad \text{i.e. } u^o \in R(f).$$

If we set $I \underset{Df}{=} \{ i \in \{1, \ldots, q\} / \; d^{i^T} u^o = 0 \}$, then

$$u^o \in \tilde{V}, \; \|u^o\| = 1, \; d^{i^T} u^o = 0, \; i \in I.$$

Since $\tilde{V}$ satisfies (MIG) and since we have $d^{i^T} u \leq 0$, for all $i \in I$ and all $u \in V$, there is a vector $v \in R^n$, $v_1, \ldots, v_s$, possessing the properties

$$v \in \tilde{V}, \; \|v\| \geq 1, \; v^T u^o \geq 1/2, \; d^{i^T} v = 0 \; \forall i \in I.$$

Now we set $y^t \underset{Df}{=} x^t - v$. Then, because of $(p^t, b^t) \in \mathcal{E}$, $t = 1, 2, \ldots$, we obtain the estimate

$$f(x^t - kv) + p^{t^T}(x^t - kv) = f(x^t) + p^{t^T} x^t - kp^{t^T} v \leq$$

$$\leq f(x^t) + p^{t^T} x^t \leq a_t \quad \forall k \geq 0.$$

Using this estimation one completes the proof in the same way as it was done in the case of Lemma 5. #

The following two theorems demonstrate that both the lower semicontinuity of the value function $\varphi$ and the upper semi-continuity of the level-set-mapping N are ensured if the cone $\tilde{V}$ satisfies the condition (MIG).

Theorem 8.

If $f(x)$ is a convex polynomial function without a linear term and $\tilde{V}$ satisfies the condition (MIG), then the value function $\varphi$ is lower semi-continuous.

Proof. If the theorem is false, then there is a point $(p^o, b^o) \in \mathcal{E}$ at which the value function is not lower semicontinuous. Then there exists a sequence $(p^t, b^t) \in \mathcal{E}$, $t = 1, 2, \ldots$, with $(p^t, b^t) \longrightarrow (p^o, b^o)$ such that, for some $\varepsilon > 0$,

$$\varphi(p^t, b^t) < \varphi(p^o, b^o) - \varepsilon \quad \forall t \geq 1. \tag{26}$$

Let $x^t \in N(p^t, b^t, \varphi(p^t, b^t) + \frac{\varepsilon}{2})$, $t = 1, 2, \ldots$, be points of minimal norm. Since the sequence $\{x^t\}$ does not possess an accumulation point, it must hold that $\|x^t\| \longrightarrow \infty$. Because of (26), this is $x^t \in N(p^t, b^t, \varphi(p^o, b^o))$.

Therefore, we can apply Lemma 7., i.e. there exists a subsequence $\{x^r\}$ and there is a vector $v \in \tilde{V}$, $v \neq 0$ and $v_1, \ldots, v_s$ integer, such that

$$x^r - v \in G(b^r), \quad \|x^r - v\| < \|x^r\|. \tag{27}$$

Since one concludes the inequality ${p^r}^T v \geq 0$ from $(p^r, b^r) \in \mathcal{E}$, we have the estimation

$$f(x^r - v) + {p^r}^T (x^r - v) = f(x^r) + {p^r}^T x^r - {p^r}^T v \leq$$

$$\leq f(x^r) + {p^r}^T x^r \leq \varphi(p^r, b^r) + \frac{\varepsilon}{2}.$$

Because of (27), this inequality gives us a contradiction to the choice of $x^t$, which proves the theorem. #

Theorem 9.

Let $f(x)$ be a convex polynomial function without a linear term and let $\tilde{V}$ satisfy the condition (MIG). Further, let

$$(p^t, b^t, a_t) \in \mathcal{E}_N, \quad t = 1, 2, \ldots ;$$

$$(p^t, b^t, a_t) \longrightarrow (p^o, b^o, a_o) \text{ with } (p^o, b^o) \in \mathcal{E} .$$

Then the level-sets possess the property

$$\forall \varepsilon > 0 \; \exists t(\varepsilon): N(p^t, b^t, a_t) \subset U_\varepsilon N(p^o, b^o, a_o) \; \forall t > t(\varepsilon).$$

Proof. We assume that the theorem is false. Then, without loss of generality* we have

$$A_t \underset{Df}{=} N(p^t, b^t, a_t) \setminus U_\varepsilon N(p^o, b^o, a_o) \neq \emptyset, \quad t = 1, 2, \ldots,$$

for some $\varepsilon > 0$. Now, we select points $x^t \in A_t$, $t = 1, 2, \ldots$, of minimal norm. This sequence $\{x^t\}$ cannot possess an accumulation point because it would have to belong to the set $N(p^o, b^o, a_o)$ implying that $x^t \in U_\varepsilon(p^o, b^o, a_o)$ would have to hold for infinitely many t. Thus, $\|x^t\| \longrightarrow \infty$. By Lemma 7., there is a vector $v \in \tilde{V}$ with $v \neq 0$ and $v_1, \ldots, v_s$ integer and there is a subsequence $\{y^r\}$ of $\{y^t\}$ defined by $y^t \underset{Df}{=} x^t - v$ which possesses the properties (i), (ii) and (iii) of Lemma 7. On account of the choice of the points $x^r$ it follows that

$$y^r \in U_\varepsilon N(p^o, b^o, a_o).$$

Let $z^r \in N(p^o, b^o, a_o)$ be chosen such that $\|y^r - z^r\| < \varepsilon$. Then, from $\|x^r - (z^r + v)\| < \varepsilon$ and $z^r + v \in G(b^o)$ we may conclude that the inequality

$$a_o < f(z^r + v) + {p^o}^T (z^r + v) = f(z^r) + {p^o}^T z^r + {p^o}^T v \leq a_o + {p^o}^T v$$

must hold and, therefore

$${p^o}^T v > 0.$$

By (iii) of Lemma 7., we have $x^r - kv \in G(b^r)$ and

$$f(x^r-kv) + p^{r^T}(x^r-kv) = f(x^r) + p^{r^T}x^r - k\cdot p^{r^T}v \leq a_r - \frac{1}{2}k\, p^{o^T}v,$$

for all $r > r(k)$, where k is an arbitrarily fixed natural number.
Now, we obtain from $a_r \longrightarrow a_o$ that $\underline{\lim}\, \varphi(p^r,b^r) = -\infty$, which contradicts $(p^o,b^o) \in \mathcal{E}$ and the fact that $\varphi$ is lower semicontinuous on $\mathcal{E}$. #

Corollary 4.

If the hypotheses of Theorem 9. are satisfied, then $N(p^o,b^o,a_o) \neq \emptyset$.

Proof. Analogously to Corollary 1. #

Corollary 4. directly allows to formulate the following existence theorem for the class of problems P(p,b) considered here.

Theorem 10.

When f(x) is a convex polynomial function without a linear term and the cone $\tilde{V}$ satisfies the condition (MIG), then the equivalence

$$P(p,b) \text{ has an optimal solution} \Longleftrightarrow (p,b) \in \mathcal{E}$$

is true.

The next theorem is related to the upper semicontinuity of the $\varepsilon$-optimal-set function $\tilde{\Psi}$.

Theorem 11.

Let f(x) be a convex polynomial function without a linear term and let $\tilde{V}$ satisfy (MIG).
If the value function $\varphi$ is continuous on $\mathcal{E}^* \subset \mathcal{E}$, then the $\varepsilon$-optimal-set function $\tilde{\Psi}$ is u.s.c. on $\mathcal{E}^* \times R_+$.

Proof. The proof runs analogously like the one of Theorem 5.; one has only to replace Theorem 4. by Theorem 9. #

If the constraint sets G(b) of the problem P(p,b) are given by (20), then, proceeding as in the previous section, we can expose such subsets $\mathcal{E}^* \subset \mathcal{E}$ on whicn the value function $\varphi$ is continuous.

Theorem 12.

Let f(x) be a convex polynomial function without a linear term. Let the constraint set G(b) given by (20) satisfy the (CPC) condition. Further, let both the cones $V = 0^+M(b)$ and $\tilde{V}$ satisfy (MIG). Finally, let $b^o \in B$ be arbitrarily fixed, $B(b^o)$ as in (5.8), and let

$$\mathcal{E}^* \underset{Df}{=} \{(p,b) \in \mathcal{E} \,/\, b \in B(b^o)\}.$$

Then

(i) The value function $\varphi$ is continuous on $\mathcal{E}^*$;

(ii) The $\varepsilon$-optimal-set function $\tilde{\Psi}$ is u.s.c. on $\mathcal{E}^* \times R_+$.

Proof. One shows (i) following the same way as in (i) of Theorem 6. The proposition (ii) follows from (i) if one takes Theorem 11. into account. #

Remark 3.

If the convex polynomial function f(x) (without a linear term) has only rational coefficients and the cone V satisfies (MIG), then the cone $\tilde{V}$ satisfies (MIG), too. This becomes clear if one takes into account that in this case the linear subspace K(f) has integer generators.

Remark 4.

If the assumptions of Theorem 12. are fulfilled, then the domain of finiteness $\mathcal{E}$ of the value function $\varphi$ coincides with the set of all pairs (p,b) for which P(p,b) has an optimal solution. By Theorem 7. the set $\mathcal{E}$ can be divided into disjoint "stability-sets" of the form $-\tilde{V}^p \times B(b^o)$. On every such stability set the value function $\varphi$ is continuous and the optimal-set function $\Psi$ is u.s.c. Moreover, the $\varepsilon$-optimal-set function $\tilde{\Psi}$ is u.s.c. on every set $-\tilde{V}^p \times B(b^o) \times R_+$.

## 7. The Existence of Optimal Points for Integer Optimization Problems

This chapter deals with the question under which conditions (mixed-) integer optimization problems possess an optimal solution.
Let the following general (mixed-) integer problem be given:

$$\inf\{f(x) \ / \ x \in M_s\} \tag{1}$$

where $M \subset R^n$, $M_s = \{x \in M \ / \ x_1, \dots, x_s \text{ integer}\}$ and $f: M \longrightarrow R$.
It is clear that the conditions

$$(NC) \quad \begin{array}{l} M_s \neq \emptyset, \\ \inf\{f(x) \mid x \in M_s\} > -\infty \end{array}$$

are necessary for the existence of an optimal solution.
Our aim is to expose such classes of problems (1) for which at most weak additional assumptions have to be imposed besides the trivial necessary conditions (NC) in order to ensure the existence of an optimal solution of (1).
Often it may turn out that checking whether the necessary conditions are fulfilled is a tough task, therefore, it will be useful to have a more tractable equivalent version of the necessary conditions (NC).
In particular, we consider cases for which the second condition in (NC) may be replaced by $\inf\{f(x) \ / \ x \in M\} > -\infty$ if $M_s \neq \emptyset$.

### 7.1. Concave Objective Functions

The following simple proposition for problems (1) with a concave (or quasiconcave) objective function is basic for deriving existence theorems for integer problems.

**Lemma 1.**
Let $M_s$ be non-empty and f a quasi-concave function on conv $M_s$. Then problem (1) has an optimal solution if and only if the problem

$$\inf\{f(x) \ / \ x \in \text{conv } M_s\} \tag{2}$$

has an optimal solution. Moreover, each optimal solution of (1) solves problem (2), too.

Proof. If $x \in \text{conv } M_s$, then, by a well-known theorem of CARATHEODORY, there are finitely many points $x^1, \dots, x^r \in M_s$ with
$x = \sum_{i=1}^{r} \alpha_i x^i$, where $\alpha_i \geq 0$ and $\sum_{i=1}^{r} \alpha_i = 1$. Since f(x) is quasiconcave on conv $M_s$, we have $f(x) \geq \min\{f(x^1), \dots, f(x^r)\}$ and, using this fact, it is easy to verify the propositions. #

Lemma 1. enables us to extend known existence results for optimization problems without integer requirements to (mixed-) integer cases; of course, one has to impose appropriate assumptions on the convex hull of the feasible set $M_s$.

Theorem 1.

The necessary conditions (NC) are sufficient for the existence of an optimal point of (1) if one of the following two assumptions is fulfilled:

(a) The convex hull conv $M_s$ is a convex polyhedron and, f is concave on it.

(b) The convex hull conv $M_s$ may be represented as the sum of a compact convex set and a closed convex cone and, f is a continuous concave function on conv $M_s$.

Proof. One obtains the propositions as consequences from Lemma 1. and known existence results for the corresponding "continuous" problems (2) (see e.g. BELOUSOV (1977)). #

Remark 1.

The importance of the required concavity of the objective function f becomes clear by considering the quasiconcave function $f(x) = e^x$. Obviously, the problem $\inf \{e^x \;/\; x \text{ integer}\}$ has no solution, because of the fact that $e^x$ is not concave on the set R of real numbers.

Lemma 2.

Let $M \subset R^n$ be represented by $M = C + V$ where C is a compact set and V a closed convex cone. Furthermore, let the (mixed-) integer points of $M_s$ be uniformly distributed in M. If now f(x) is a continuous concave function on conv C + V, then the equivalence

$$\inf\{f(x) \;/\; x \in M_s\} > -\infty \iff \inf\{f(x) \;/\; x \in M\} > -\infty$$

holds.

Proof. We have only to show that the implication

$$\inf\{f(x) \;/\; x \in M\} = -\infty \implies \inf\{f(x) \;/\; x \in M_s\} = -\infty \qquad (3')$$

is valid. If the left-hand side of (3') is true, then, because of the fact that f is a continuous concave function on conv C + V, we know from convex analysis that there is a vector $u \in V$ for which

$$\inf\{f(x) \;/\; x = \bar{x} + tu,\; t \geq 0\} = -\infty \qquad (4')$$

holds for any fixed point $\bar{x} \in$ conv C + V. Let us consider such a vector u and an $\bar{x} = c \in M$.

Since the points of $M_s$ are uniformly distributed in M, we have that

$$\delta \underset{Df}{=} \sup\{d(x, M_s) \;/\; x \in M\} < \infty.$$

For the sequence $\{x^t\} \subset M$ given by $x^t = c + tu$, $t = 1,2,\ldots$, there exists a sequence $\{y^t\} \subset M_s$ such that

$$\|x^t - y^t\| < \delta + 1, \; t = 1,2,\ldots .$$

We show now that

$$\inf\{f(y^t) \ / \ t = 1,2,\ldots\} = -\infty$$

holds. If this were not the case, then we could conclude that

$$\inf\{f(y) \ / \ y \in Y\} > -\infty$$

would be true, where Y denotes the unbounded set $\operatorname{conv}\{y^t \ / \ t = 1,2,\ldots\}$. By the construction of the sequence $\{y^t\}$, the set Y had to contain a half-line of the form $x = c^* + tu$, $t \geq 0$, where $c^* \in M$ and u as before. This would imply

$$\inf\{f(x) \ / \ x = c^* + tu, \ t \geq 0\} > -\infty,$$

contradicting (4'). #

The following example shows that the continuity of the function f is necessary for the validity of Lemma 2.:

Example 1.

The function f given by

$$f(x) \underset{Df}{=} \begin{cases} 0 & \text{if } x_1 \neq 1/2 \\ -x_2 & \text{otherwise,} \end{cases}$$

is concave on the set

$$M = \{x \in R^2 \ / \ 0 \leq x_1 \leq 1/2, \ x_2 \geq 0\}$$

but not continuous. Here one has

$$\inf\{f(x) \ / \ x \in M, \ x_1, x_2 \text{ integer}\} = 0$$

and,

$$\inf\{f(x) \ / \ x \in M\} = -\infty.$$

## 7.2. Polynomial Objective Functions

In Chapter 6. we already derived existence theorems for quadratic and polynomial convex objective functions, for these cases the reader is referred to the corresponding sections. However, we can extend the result valid for convex polynomial objective functions to the case of quasiconvex polynomial objective functions.

We consider a quasiconvex polynomial function f(x) given in the form $f(x) = q_m(x) + \ldots + q_1(x) + q_0$, where each function $q_i(x)$ is a form of degree i. Corresponding to Lemma 4.4. we use the index $i_0$ in order to define the linear subspace

$$K^*(f) \underset{Df}{=} \{u \in R^n \ / \ q_m(u) = 0, \ldots, q_{i_0+1}(u) = 0\}.$$

If all coefficients of the polynomials appearing in the description of $K^*(f)$ are rational numbers, then we know (by section 4.1.) that the linear subspace can be integerly generated, i.e. it satisfies (IG) (see Chapter 5.). Furthermore, if K(f) (defined in Lemma 4.4.) is also a linear subspace, then it coincides with $K^*(f)$. If, in particular, f is a convex polynomial function without a linear term, we have $K^*(f) = K(f) = R(f)$ (cf. Lemma 4.4.).

Analogously as in the case of a convex polynomial function (see (6.21)) we define the set

$$\tilde{V} \underset{Df}{=} K^*(f) \cap V \qquad (5')$$

for a quasiconvex polynomial function f(x); where V is a polyhedral convex cone.

Theorem 2.

If f is a quasiconvex polynomial function and M given by $M = C + V \subset R^n$ where C is a compact set and V a polyhedral convex cone, then the following propositions are true:

(i) If V satisfies (MIG), $M_s \neq \emptyset$ and $\inf\{f(x) \ / \ x \in M\} = \infty$, then $\inf\{f(x) \ / \ x \in M_s\} = -\infty$ holds, too.

(ii) If the cone $\tilde{V}$ from (5') satisfies (MIG), then the optimization problem (1) has an optimal solution if and only if the conditions (NC) are satisfied.

Proof. For shortness, we only sketch the proof. We proceed analogously as in section 6.3., i.e. one has to modify the Lemmata 6.6. and 6.7. (here, in particular, the propositions (i) and (ii)) and Corollary 6.3. Thereby, one has to take into account the results of section 4.1. In particular, one has to demonstrate that the conditions (NC) are sufficient for the existence of an optimal solution of (1).

This can be done analogously as in the proof of Theorem 6.2. Finally, we put $a_o \underset{Df}{=} \inf\{f(x) \ / \ x \in M_s\}$ and $a_t = a_o + \frac{1}{t}$, $t = 1,2,\ldots,$ and define $N(\alpha) \underset{Df}{=} \{x \in M_s \ / \ f(x) \leq \alpha\}$. Selecting points $x^t$ of minimal norm from $N(a_t)$ one concludes that the sequence $\{x^t\}$ must have an accumulation point $x^o$ belonging to $N(a_o)$. #

Example 2.

The integer optimization problem

$$\inf\{(\sqrt{2}\,x_1 - x_2)^2 \ / \ x_1 \geq 1, \ x_1, x_2 \text{ integer}\}$$

belongs to the class of problems considered in the Theorem 2. and 4.2. But, on the one hand we have that

$$\inf\{(\sqrt{2}\,x_1 - x_2)^2 \ / \ x_1 \geq 1, \ x_1, x_2 \text{ integer}\} = 0$$

and, on the other hand that

$$(\sqrt{2}\,x_1 - x_2)^2 > 0 \text{ if } x_1 \geq 1 \text{ and } x_1, x_2 \text{ integer},$$

because

$$\tilde{V} = \{u \in R^2 \,/\, \sqrt{2}\, u_1 - u_2 = 0,\ u_1 \geq 0\}$$

does not satisfy the condition (IG) required in both theorems.

ANDRONOW et al. (1982) have shown that a polynomial function f of degree three attains its infimum on a non-empty convex polyhedron if f is bounded from below on this polyhedron. A corresponding statement for mixed-integer problems will not be true even if all coefficients of the objective function and the constraints are rational numbers. The following example due to ANDRONOW et al. (1982) illustrates the situation:

Example 3.

The mixed-integer problem $\inf\{f(x) \,/\, x \in M_2\}$ is considered where

$$f(x_1, x_2, x_3) = x_1^3 + 2x_2^3 + 4x_3^3 - 6x_1x_2x_3,$$

$$M_2 = \{x \in R^3 \,/\, x_i \geq 1,\ i = 1,2,3,\ x_1, x_2 \text{ integer}\}.$$

This problem has no optimal solution since

$$\inf\{f(x) \,/\, x \in M_2\} = 0 \text{ and, } f(x) > 0\ \forall x \in M_2,$$

which one can verify using the well-known relation between the geometrical and the arithmetical mean.

Quasiconcave polynomial objective functions allow to derive an existence-theorem similar to Theorem 1. (b).

Theorem 3.

If f(x) is a quasiconcave polynomial function and the convex hull conv $M_s$ of the feasible region $M_s$ possesses a representation as the sum of a compact set C and a closed convex cone V, then the (mixed-) integer problem (1) has an optimal solution if and only if the conditions (NC) are satisfied.

Proof. On account of Lemma 1., it suffices to show that problem (2) has an optimal solution if

$$\inf\{f(x) \,/\, x \in \text{conv } M_s\} > -\infty.$$

Since f(x) is a quasiconcave polynomial function and it is bounded from below on every half-line contained in conv $M_s$, one obtains, by conv $M_s = C + V$, that

$$\inf\{f(x) \,/\, x \in \text{conv } M_s\} = \inf\{f(x) \,/\, x \in C\}$$

holds. We have $C \subset C + V$ because V is a closed cone and, therefore, every solution of the (trivially) solvable problem $\inf\{f(x) \,/\, x \in C\}$ is also a solution of (2). #

With regard to the condition

$$\inf\{f(x) \;/\; x \in M_s\} > -\infty$$

we prove the following lemma:

Lemma 3.

Let M = C + V be true, where C is a bounded set and V is a convex cone. Furthermore, let f(x) be a quasiconcave polynomial function. Under the assumption that the points of $M_s$ are uniformly distributed in M one has the following equivalence:

$$\inf\{f(x) \;/\; x \in M_s\} > -\infty \Longleftrightarrow \inf\{f(x) \;/\; x \in M\} > -\infty.$$

Proof. Obviously, it suffices to show that the implication

$$\inf\{f(x) \;/\; x \in M\} = -\infty \Longrightarrow \inf\{f(x) \;/\; x \in M_s\} = -\infty$$

is true. Let $\inf\{f(x) \;/\; x \in M\} = -\infty$ be. Since f(x) is a quasiconcave polynomial function, it follows from

$$\inf\{f(x) \;/\; x = c + tv,\; t > 0\} > -\infty$$

that

$$f(c) \leq \inf\{f(x) \;/\; x = c + tv,\; t > 0\}.$$

Hence, there must be a point $\bar{c} \in C$ and a vector $\bar{v} \in V$ such that

$$\inf\{f(x) \;/\; x = \bar{c} + t\bar{v},\; t > 0\} = -\infty$$

holds. Now, let $\delta > 0$ be chosen such that

$$\delta > \sup\{d(x, M_s) \;/\; x \in M\}$$

and, let $\{t_k\}$ be a sequence with $t_k \longrightarrow \infty$. Corresponding to this sequence we select points $x^k \in M_s$ for which $\|x^k - (\bar{c} + t_k\bar{v})\| \leq \delta$. Because of the fact that f(x) is quasiconcave one concludes, applying Lemma 4.1., that $f(x^k) \longrightarrow -\infty$. #

The next existence theorem is related to a (mixed-) integer problem (1) having a quasiconvex polynomial objective function f(x) and a constraint set M represented as the solution set of a finite system of quasiconvex polynomial inequalities.

Theorem 4.

Let f(x) as well as $p_1(x), \ldots, p_m(x)$ be quasiconvex polynomial functions and, let the sets

$$M \underset{Df}{=} \{x \in R^n \;/\; p_i(x) \leq 0,\; i = 1, \ldots, m\}$$

and

$$\hat{M} \underset{Df}{=} \{x \in R^n \;/\; p_i(x) \leq p_i(0),\; i = 1, \ldots, m;\; f(x) \leq f(0)\}$$

be defined.

If the recession cone $\hat{V}_{st}$ of the stable set $\hat{M}_{st}$ (see(4.21)) corresponding to $\hat{M}$ satisfies (MIG), then the two following statements hold:

(i) Problem (1) has an optimal solution if and only if (NC) is fulfilled.

(ii) If $M_s \neq \emptyset$, then

$$\inf\{f(x) \;/\; x \in M_s\} > -\infty \Longleftrightarrow \inf\{f(x) \;/\; x \in M\} > -\infty.$$

Proof. We set

$$G(b) \underset{Df}{=} \{x \in R^n \;/\; p_i(x) \leq 0,\; i = 1,\ldots,m;\; f(x) \leq b,\; x_1,\ldots,x_s \text{ integer}\}.$$

(i) Let us define

$$b_o \underset{Df}{=} \inf\{f(x) \;/\; x \in M_s\},$$

$$b_t \underset{Df}{=} b_o + \frac{1}{t}\,,\quad t = 1,2,\ldots\,.$$

Then, obviously, $G(b_t) \neq \emptyset$, $t = 1,2,\ldots$, and, by Theorem 5.3. (ii), $G(b_o) \neq \emptyset$.

(ii) If $\inf\{f(x) \;/\; x \in M\} = -\infty$, then we have, by Theorem 5.3., that $G(b) \neq \emptyset\ \forall b \in \mathbb{R}$, i.e. $\inf\{f(x) \;/\; x \in M_s\} = -\infty$. #

One immediately sees the importance of the (MIG) condition for the recession cone $\hat{V}_{st}$ considering once more Example 2. The problem

$$\inf\{(\sqrt{2}\,x_1 - x_2)^2 \;/\; x_1 \geq 1,\; x_1, x_2 \text{ integer}\}$$

has no optimal solution and, we have

$$\hat{M} = \{x \in R^n \;/\; -x_1 \leq 0,\; (\sqrt{2}\,x_1 - x_2)^2 \leq 0\};$$

$$\hat{M}_{st} = \{x \in R^n \;/\; (\sqrt{2}\,x_1 - x_2)^2 \leq 0\},$$

$$\hat{V}_{st} = \{u \in R^n \;/\; \sqrt{2}\,u_1 - u_2 = 0\}.$$

Obviously, the linear subspace $\hat{V}_{st}$ does not possess an integer generator.

Remark 2.

If all the quasiconvex polynomial functions needed to describe $\hat{M}_{st}$ have only rational coefficients, then the cone $\hat{V}_{st}$ satisfies (IG), which directly follows from Lemma 4.6.

Remark 3.

If $f(x)$ is a continuous convex function on a closed convex set $M \subset R^n$, then all level sets $N(\alpha) \underset{Df}{=} \{x \in M \;/\; f(x) \leq \alpha\}$ are empty or compact if this is the case at least for some $\alpha$. In such a case it is clear that problem (1) has an optimal solution if $M_s \neq \emptyset$ holds (see also Corollary 3.2.).

## 7.3. Some Special Mixed-Integer Optimization Problems.

This section contains an extension of a simple existence theorem for pure integer problems to mixed-integer problems.

Let $T \subset \{x \in R^n \ / \ x_1, \ldots, x_n \text{ integer}\}$. Then a vector-valued function $F: T \longrightarrow R^m$ is said to be quasi-integer valued on T if there is a natural number $\alpha \neq 0$ such that

$$\alpha \cdot F(x) \in \{v \in R^m \ / \ v_1, \ldots, v_m \text{ integer}\} \ \forall x \in T.$$

For example, a polynomial function p(x) with rational coefficients is a quasi-integer valued function.

Remark 4.

Let $f^{\cdot}: t \longrightarrow R$ be a quasi-integer valued function on $T \neq \emptyset$ and, let $\inf\{f(x) \ / \ x \in T\} > -\infty$ hold. Then the problem $\inf\{f(x) \ / \ x \in T\}$ has an optimal solution.

Based on the statement contained in Remark 4. we now derive existence theorems for special mixed-integer optimization problems. For that purpose we consider the following general representation of a mixed-integer problem

$$\inf\{f(z,y) \ / \ (z,y) \in G\} \tag{3}$$

where

$$G \subset \{(z,y) \in R^n \ / \ z \in R^s, \ z \text{ integer}, \ y \in R^{n-s}\}.$$

First, we shall explain the general idea and, for that we define

$$G(\lambda) \underset{Df}{=} \{y \in R^{n-s} \ / \ (\lambda, y) \in G\}, \ \lambda \in R^s,$$

$$B \underset{Df}{=} \{\lambda \in R^s \ / \ \lambda \text{ integer}, \ G(\lambda) \neq \emptyset\}$$

and the value function $\varphi: R^s \longrightarrow R$ by

$$\varphi(\lambda) \underset{Df}{=} \inf\{f(\lambda,y) \ / \ y \in G(\lambda)\}.$$

Remark 5. (The partitioning principle)

The pair $(z^0, y^0)$ is an optimal solution of the problem (3) if and only if the point $z^0$ is an optimal solution of the pure integer problem

$$\inf \ \{\varphi(\lambda) \ / \lambda \in B\} \tag{4}$$

and, $y^0$ is an optimal solution of the parametric problem

$$\inf\{f(\lambda,y) \ / \ y \in G(\lambda)\}, \tag{5}$$

for $\lambda = z^0$ (The proof is quite technical).

Remark 6.

If the objective function $\varphi(\lambda)$ of (4) is quasi-integer valued on B and (5) has an optimal solution for every $\lambda \in B$, then one immediately obtains from Remark 4. a proposition on the existence of an optimal point in the problem (3).

The considerations carried out in the Remarks 5. and 6. now shall be used in order to derive an existence-theorem for problems of the type (3) when the constraint set G is given by

$$G = \{(z,y) \in R^n \ / \ F(z) + Ay \leq 0, \ z \in Y\} \tag{6}$$

where A denotes a rational (m,n-s)-matrix, Y is a subset of $T \underset{Df}{=} \{z \in R^s \ / \ z \text{ integer}\}$ and $F: T \longrightarrow R^m$ is a quasi-integer valued vector-function.

Theorem 5.

Let the constraint set G, given by (6), be non-empty and let $\inf\{f(z,y) \ / \ (z,y) \in G\} > -\infty$. Then problem (3) has an optimal solution if the objective function f(z,y) has one of the properties (a) or (b):

(a) $f(z,.)$ is a concave function on the convex polyhedron G(z) for every $z \in B$, and, for every integer $d > 0$, there is an integer $\alpha(d) > 0$ such that $\alpha(d) \cdot f(z, \frac{1}{d}y)$ is integer, for all integer vectors $y \in R^{n-s}$ and all $z \in B$.

(b) $f(z,y) = f_1(z) + z^T Hy + f_2(y)$

where $f_1(z)$ is quasi-integer valued for every integer vector $z \in R^s$ and, $f_2(y)$ is a quadratic function having rational coefficients and H is a rational (s,n-s)-matrix.

Proof. From $G \neq \emptyset$ and $\inf\{f(z,y) \ / \ (z,y) \in G\} > -\infty$ it follows that $B \neq \emptyset$ and $\inf \{\varphi(\lambda) \ / \ \lambda \in B\} > -\infty$.

Furthermore, it is clear, that a quadratic (or a concave) function attains its infimum on a non-empty convex polyhedron if this function is bounded from below on it. Therefore, with respect to Remark 6., it suffices in both cases (a) or (b) to show that the function $\varphi(\lambda)$ is quasi-integer valued on B.

Case (a). Let $\lambda \in B$. Since the function $f(\lambda,.)$ is concave on the polyhedron $G(\lambda)$, the optimal set of (5) contains a smallest-dimensional face $S(\lambda)$ of $G(\lambda)$. Because A is a rational matrix and, since F is quasi-integer valued on T, one easily sees that all the faces $S(\lambda)$, $\lambda \in B$, must contain points $y(\lambda)$, which are of the form

$$y(\lambda) = \frac{1}{d} y^*(\lambda), \tag{7}$$

where $y^*(\lambda)$ is an integer vector and, d is a natural number independent of $\lambda$.

Hence, the function $\varphi(\lambda)$ may be represented as

$$\varphi(\lambda) = f(\lambda, \frac{1}{d} y^*(\lambda)), \ \lambda \in B,$$

and now, taking the property (a) into account, one concludes that $\varphi(\lambda)$ is quasi-integer valued on B.

Case (b). We show that problem (5) has also in this case solutions of the form (7) if $\lambda \in B$. The function $\varphi(\lambda)$ then possesses the representation

$$\varphi(\lambda) = f_1(\lambda) + \lambda^T H(\tfrac{1}{d}y^*(\lambda)) + f_2(\tfrac{1}{d}y^*(\lambda)),\ \lambda \in B,$$

and one concludes, using the property (b) of $f(z,y)$, that $\varphi(\lambda)$ is quasi-integer valued on B. It remains to show that there are solutions $y(\lambda)$, $\lambda \in B$, of the form (7).

We assume without loss of generality that in (6) the condition $y \geqq 0$ is included and we write

$$\bar{F}(\lambda) + \bar{A}y \leqq 0,\ y \geqq 0$$

instead of $F(\lambda) + Ay \leqq 0$. Further, there are a rational $(n-s,n-s)$-matrix D and a rational vector $p \in R^{n-s}$ such that

$$f_2(y) = y^T D y + p^T y$$

may be written. For an arbitrary $\lambda \in B$ the Kuhn-Tucker-system of (5) has the form

$$\begin{pmatrix} u \\ v \end{pmatrix} + \begin{pmatrix} D+D^T & \bar{A}^T \\ -\bar{A} & 0 \end{pmatrix} \begin{pmatrix} y \\ t \end{pmatrix} = \begin{pmatrix} p+H^T\lambda \\ -\bar{F}(\lambda) \end{pmatrix}, \tag{8}$$

$$u \geqq 0,\ v \geqq 0,\ y \geqq 0,\ t \geqq 0, \qquad u^T y = 0,\ v^T t = 0. \tag{9}$$

With respect to Lemma 6.1., the polyhedron $W(\lambda) \subset R^{2(m-s)}$, $\lambda \in B$, defined by (8) and (9) possesses a vertex point

$$\bar{e}(\lambda)^T = (\bar{u}(\lambda)^T, \bar{v}(\lambda)^T, \bar{y}(\lambda)^T, \bar{t}(\lambda)^T)$$

such that $\bar{y}(\lambda)$ solves problem (5). Because of the rationality of the involved matrices D, $\bar{A}$, H and p, and since the vector function $\bar{F}(\lambda)$ is quasi-integer valued, every vertex point $e(\lambda)$ of the polyhedron $W(\lambda)$, $\lambda \in B$, may be represented in the form

$$e(\lambda) = \frac{1}{d}\, e^*(\lambda)$$

where $e^*(\lambda)$ is an integer vector in $R^{2(m-s)}$ and, d is a natural number independent of $\lambda$ . #

Example 4.

The following two functions have the property (a) of Theorem 5:

1. $f(z,y) = f_1(z) + z^T D y + f_2(y)$

   where $f_2(y)$ is a concave polynomial function with rational coefficients, D is a rational $(s,n-s)$-matrix and, $f_2(y)$ is quasi-integer valued for integer vectors $y \in R^{n-s}$.

2. $f(z,y) = h(z) + \sum_{i=1}^{s} (z_i)^2 f_i(y)$

   where $h(z)$ is quasi-integer valued on the set of integer vectors in

$R^s$ and, $f_1(y),\dots,f_s(y)$ are concave polynomial functions with rational coefficients.

Example 5.

Integer quadratic optimization problems yield a particular class of problems treated in Theorem 5.(b) if all the coefficients of the quadratic objective function as well as of the linear constraints are rational numbers. (From Theorem 6.2. we already know the existence statement for this case included in Theorem 5.(b).)

Remark 7.

The proof of Theorem 5. permits a generalization of a result due to WOLSEY (1971), giving a reduction of mixed-integer linear programs to pure integer linear programs (if all data are rational). Instead of the problems (3), (6) treated in Theorem 5. one can solve the pure integer problem

$$\inf\left\{f(z,\tfrac{1}{d}w) \;/\; d\cdot F(z) + Aw \leq 0,\; z\in Y,\; w \text{ integer}\right\}$$

where d is a natural number characterized in the proof of Theorem 5. for which WOLSEY's discussion may be applied. Now, if $(z,w)$ is an optimal point of the latter pure integer problem, then $(z,y)$ with $y = \frac{1}{d}w$ solves the original mixed-integer problem.

## 8. Quantitative Stability of (Mixed-) Integer Linear Optimization Problems

The aim of this chapter consists in the presentation of quantitative stability results for parametric (mixed-) integer linear optimization problems, where quantitative stability is understood in terms of estimates for the value function and the optimal set-function and their quasi-Lipschitzian behaviour (see Chapter 5.); hereby, a quantitative characterization of the relations between the (mixed-) integer problems and the associated linear programming relaxations is also a natural and important question.
Throughout this chapter we deal with the interesting case in which the right-hand sides of the linear constraints are the only parameters. The treatments of Chapter 5. allow a unified approach for this case. The theorems below represent an extension of the comprehensive analysis due to BLAIR/JEROSLOW (1977, 1979). The importance of such a quantitative concept is shown in Chapter 9. where it will be applicated.

Let us consider the problem

$$P(b) \quad \inf \{ c^T z + d^T y \;/\; x = (z,y) \in G(b) \} \;,\; b \in R^m,$$

where the constraint sets G(b) are now given in the form

$$G(b) \underset{Df}{=} \{ x \in R^n \;/\; x = (z,y),\; Az+Qy = b;\; z,y \geqq 0,\; z \in R^s \text{ integer} \} \qquad (1)$$

and the general assumption:

$$D \underset{Df}{=} (A,Q) \text{ is an integer matrix.}$$

For shortness, we write G(b) also in the form

$$G(b) = \{ x \in R^n \;/\; Dx = b,\; x \geqq 0,\; x_1,\dots,x_s \text{ integer} \}.$$

The recession cone of the polyhedron given by $Dx = b$ and $x \geqq 0$ is

$$V = \{ x \in R^n \;/\; x = (z,y),\; Az+Qy = 0,\; z,y \geqq 0 \}.$$

The integer-assumption above as well as the representation of the constraint sets G(b) by equations is made for convenience and in view of a unified approach. Often it would already suffice to impose the (MIG) condition for the recession cone V corresponding to G(b), which one easily verifies.
If $\varphi(b)$ denotes the value function of P(b), then the assumption "$\varphi(0) = 0$" ensures that P(b) is solvable for all $b \in B$ (=dom G).

Theorem 1.
Let $\varphi(0) = 0$. Then

(i) the optimal set-function $\psi$ is quasi-Lipschitzian on B; moreover, there is a quasi-Lipschitzian compact-valued multifunction

$\hat{K} : B \longrightarrow 2^{R^n}$ such that

$$\psi(b) = \hat{K}(b) + \hat{V}_s \quad \forall b \in B$$

where

$$\hat{V}_s \underset{Df}{=} \{x \in R^n \;/\; Dx = 0 \quad (c,d)^T x \leq 0, \; x \geq 0, \; x_1, \ldots, x_s \text{ integer}\} ;$$

(ii) there are constant numbers $\gamma > 0$ and $\delta > 0$ such that

$$|\varphi(b) - \varphi(b^*)| \leq \gamma \| b - b^* \| + \delta \quad \forall b, b^* \in B.$$

(If $s = 0$, i.e. the "continuous" problem $P^0(b)$ is considered, then $\psi$ and $\hat{K}$ are Lipschitzian and, $\delta = 0$ in (ii).)

Proof.

(i) We start showing the first part of (i), which will be used to demonstrate proposition (ii) and, both enables us to verify the remaining part of (i). By assumption, we have $\psi(b) \neq \emptyset$, for all $b \in B$. Following the technique used in the proof of Corollary 4.3. we define

$$\tilde{G}(\beta) \underset{Df}{=} \{(y,u) \in R^{2n} \;/\; Dy - Du = \beta, \; y,u \geq 0, \; y_1, \ldots, y_s, \; u_1, \ldots, u_s \text{ integer}\}$$

and

$$\tilde{V}_{2s} \underset{Df}{=} \{(y,u) \in R^{2n} \;/\; Dy - Du = 0, \; y,u \geq 0, \; y_1, \ldots, y_s, \; u_1, \ldots, u_s \text{ integer}\} .$$

Then, by Theorem 5.2., there is a compact-valued multifunction $\tilde{K}$ and, by Theorem 5.6. (iv), there are constant numbers $\rho > 0$ and $\tau > 0$ such that

$$\tilde{G}(\beta) = \tilde{K}(\beta) + \tilde{V}_{2s}, \quad \beta \in \operatorname{dom} \tilde{G},$$

$$\|(y,u)\| \leq \rho \|\beta\| + \tau, \quad \forall (y,u) \in \tilde{K}(\beta).$$

Let $b, b^* \in B$ and $x \in \psi(b)$, $x' \in \psi(b^*)$ be arbitrarily fixed. Then

$$(x,x') \in \tilde{G}(b-b^*)$$

and therefore, one finds points

$$(y,u) \in \tilde{K}(b-b^*), \quad (\bar{y},\bar{u}) \in \tilde{V}_{2s}$$

such that

$$(x,x') = (y,u) + (\bar{y},\bar{u}),$$
$$\|(y,u)\| \leq \rho \|b-b^*\| + \tau .$$

We show that $\bar{y} \in \psi(D\bar{y})$. Obviously, $\bar{y} \in G(D\bar{y})$. Assume that there is a point $z \in G(D\bar{y})$ satisfying $c^T z < c^T \bar{y}$. Then we have

$$c^T(y+z) < c^T(y+\bar{y}) = c^T x.$$

However, this contradicts $x \in \psi(b)$ because of

$y + z \geq 0$ and $y_i + z_i$, $i = 1,\ldots,s$, integer

and

$$D(y+z) = Dy + Dz = Dy + D\bar{y} = Dx = b,$$

i.e. $y + z \in G(b)$.

One analogously verifies that $\bar{u} \in \Psi(D\bar{u})$.

Since $D\bar{y} = D\bar{u}$, we have

$$c^T\bar{y} = c^T\bar{u}$$

and, for the point

$$x^* \underset{Df}{=} u + \bar{y}$$

we obtain

$$c^Tx^* = c^Tu + c^T\bar{y} = c^Tu + c^T\bar{u} = c^Tx'.$$

Because of $x^* \in G(b^*)$ (which one easily sees) this implies $x^* \in \Psi(b^*)$. Estimating

$$\|x-x^*\| = \|y+\bar{y}\|- u-\bar{y}\| = \|y-u\| \leq 2\|(y,u)\| \leq$$
$$\leq 2(\varrho\|b-b^*\| + \tau)$$

and setting $\alpha = 2\varrho$ and $\beta = 2\tau$ one obtains the first part of proposition (i).

(ii) Using the above result and notations one estimates

$$|\varphi(b) - \varphi(b^*)| = |c^T(x-x^*)| \leq \|c\|\|x-x^*\| \leq$$
$$\leq \|c\|(\alpha\|b-b^*\| + \beta)$$

and, by setting $\gamma = \alpha\|c\|$ and $\delta = \beta\|c\|$ the desired proposition (ii) follows.

Now we proceed with the proof of the second part of proposition (i). Let

$$N(b,a) \underset{Df}{=} \{x \in R^n \,/\, Dx = b,\ (c,d)^Tx \leq a,\ x \geq 0,\ x_1,\ldots,x_s \text{ integer}\}$$

and

$$\hat{V} \underset{Df}{=} \{x \in R^n \,/\, Dx = 0,\ (c,d)^Tx \leq 0,\ x \geq 0\}.$$

By the rationality of D, the polyhedral cone

$$V = \{x \in R^n \,/\, Dx = 0,\ x \geq 0\}$$

has integer generators. Since $\varphi(0) = 0$,

$$(c,d)^Tv \geq 0 \quad \forall v \in V$$

must hold and, therefore, also the polyhedral cone $\hat{V}$ has a system of integer generators (independent of the fact whether $(c,d)$ is a rational vector or not).

On this basis following a similar way as used in the proof of Theorem 5.7. one can show that there is a quasi-Lipschitzian compact-valued multifunction $\bar{K}$: dom $N \longrightarrow 2^{R^n}$ such that $N(b,a) = \bar{K}(b,a) + \hat{V}_s$, for all

$(b,a) \in \text{dom } N$. Thereby, one can use Theorem 5.1. instead of Lemma 5.4. because the set

$$\hat{M} \underset{Df}{=} \{(x,b,a) \in R^{n+m+1} \; / \; Dx-b = 0, \; (c,d)^T x-a \leq 0, \; x \geq 0\}$$

has a relatively interior point $(x,b,a)$ with $x_1,\dots,x_s$ integer, and the affine hull

$$\text{aff } \hat{M} = \text{aff } \{(x,b,a) \in R^{n+m+1} \; / \; Dx-b = 0, \; x \geq 0\}$$

has an integer basis. Let $\alpha > 0$ and $\beta > 0$ be the corresponding constant numbers such that

$$d(\overline{K}(b,a), \; \overline{K}(b^*,a^*)) \leq \alpha \|(b,a) - (b^*,a^*)\| + \beta \quad \forall (b,a), \; (b^*,a^*) \in \text{dom } N.$$

If one chooses $a = \varphi(b)$ and defines

$$\hat{K}(b) \underset{Df}{=} \overline{K}(b, \varphi(b)),$$

then one has

$$\psi(b) = \hat{K}(b) + \hat{V}_s \quad \forall b \in B$$

and, taking (ii) into account, one gets

$$\begin{aligned} d(\hat{K}(b),\hat{K}(b^*)) &\leq \alpha \|(b, \varphi(b)) - (b^*, \varphi(b^*))\| + \beta \leq \\ &\leq \alpha(\|b-b^*\| + |\varphi(b)-\varphi(b^*)|) + \beta \leq \\ &\leq \alpha(\|b-b^*\| + \gamma\|b-b^*\| + \delta) + \beta = \\ &= (\alpha+\gamma)\|b-b^*\| + \alpha\delta + \beta), \end{aligned}$$

which completes the proof of (i).
If $s = 0$, one has only to use Corollary 4.2. instead of Theorem 4.6.(iv).

#

The following Theorem 2. is devoted to relations between the (mixed-) integer problems $P(b)$ and their linear programming relaxations.
Let again $P^0(b)$ denote the parametric problem which appears when the integer requirements in $P(b)$ are omitted (i.e. $s = 0$). By $\varphi^0(b)$ and $\psi^0(b)$ we denote the corresponding value function and the optimal-set function of $P^0(b)$.

Theorem 2.

Let $\varphi(0) = 0$. Then

(i) there is a constant number $\Delta > 0$ such that

$$d(\psi(b), \psi^0(b)) \leq \Delta \quad \forall b \in B;$$

(ii) $\sup \{|\varphi(b) - \varphi^0(b)| \mid b \in B\} < \infty$;

(iii) $\text{epi } \varphi^0 = \text{conv epi } \varphi$.

Proof.

(i) Let $x^o=(z^o,y^o)\in\psi^o(b)$ for an arbitrarily chosen $b\in B$. We consider the point $x^* \underset{Df}{=} ([z^o],y^o)$ and show that $x^*\in\psi(b^*)$ for $b^* = A[z^o]+ Qy^o$. Assume that there is a point $\tilde{x} = (\tilde{z},\tilde{y})\in\psi^o(b^*)$ satisfying

$$c^T\tilde{z} + d^T\tilde{y} < c^T[z^o]+ d^Ty^o,$$

from which one obtains

$$c^T(\tilde{z} + (z^o -[z^o])) + d^T\tilde{y} < c^Tz^o + d^Ty^o.$$

On the other hand, we have

$$\begin{aligned} A(\tilde{z} + (z^o -[z^o])) + Q\tilde{y} &= A\tilde{z} + Q\tilde{y} + A(z^o -[z^o]) = \\ &= b^* + A(z^o -[z^o]) = \\ &= Az^o + Qy^o = b \end{aligned}$$

and

$$\tilde{z} + (z^o -[z^o]) \geq 0.$$

Therefore, the point $(\tilde{z} + (z^o -[z^o]),\tilde{y})$ is feasible for the problem $P^o(b)$ and has a smaller objective function value than $x^o = (z^o,y^o)$, which is impossible. Hence, $x^*\in\psi^o(b^*)$ and by $[z^o]\in Z^s$ it follows that $x^*\in\psi(b^*)$. If $\alpha$ and $\beta$ are the constant numbers of Theorem 1., then there is a point $x^1 = (z^1,y^1)\in\psi(b)$ such that

$$\|x^1-x^*\| \leq\alpha\|b-b^*\| + \beta.$$

We get

$$\begin{aligned} \|x^1-x^o\| &\leq \|x^1-x^*\| + \|x^*-x^o\| \leq\alpha\|b-A[z^o]+ Qy^o\| + \beta + \\ &+ \|([z^o]- z^o,0)\| \leq\alpha\|Az^o+Qy^o - A[z^o]+ Qy^o\| + \beta + 1 = \\ &=\alpha\|A(z^o -[z^o])\| + \beta + 1\leq \\ &\leq\alpha\|A\|\,\|z^o -[z^o]\| + \beta + 1 \leq\alpha\|A\| + \beta + 1. \end{aligned}$$

Setting $\Delta \underset{Df}{=} \alpha\|A\| + \beta + 1$, proposition (i) follows.

(ii) Let $x\in\psi(b)$ and $x^1\in\psi^o(b)$ such that $\|x-x^1\| \leq\Delta$ . Then

$$|\varphi(b) - \varphi^o(b)| =|(c,d)^T(x-x^1)| \leq\|(c,d)\|\Delta < \infty.$$

(iii) Let

$$M(b) \underset{Df}{=} \{x\in R^n \,/\, Dx = b,\ x \geq 0\}$$

be the feasible set of $P^o(b)$. Then the dual problem $D^o(b)$ has the feasible set

$$W \underset{Df}{=} \{w\in R^m \,/\, D^Tw \leq \binom{c}{d}\}.$$

We know that $B^o \underset{Df}{=}$ dom M is a polyhedral cone with a vertex point at $0\in R^m$ and, by the rationality of D, it has integer generators $b^1,\dots,b^r$. From linear parametric programming it is known that there is a finite number of vectors $w^1,\dots,w^q\in W$ such that

$$\varphi^{o}(b) = \max_{1 \leq j \leq q} b^T w^j$$

and, therefore

$$\text{epi } \varphi^{o} = \{(b,\mu) \in R^{m+1} \;/\; b \in B^{o},\ \mu \geq \varphi^{o}(b)\}$$

is a polyhedral cone with a vertex point at $0 \in R^m$.
Since $\varphi(b) \geq \varphi^{o}(b)$ and $B^{o} \supset B$, we only have to show that

$$\text{epi}\,\varphi^{o} \subset \text{conv epi}\,\varphi \tag{2}$$

holds.
Let $(b,\mu) \in \text{epi}\,\varphi^{o}$. Then there are numbers $\alpha_i \geq 0$, $i = 1,\dots,r$, such that $b = \sum_{i=1}^{r} \alpha_i b^i$, and it holds that $\mu \geq \varphi^{o}(b)$. One easily sees that $(kb^i, k\varphi(b^i)) \in \text{epi}\,\varphi$ (analogously defined as epi $\varphi^{o}$), for all integers $k \geq 0$ and all $i = 1,\dots,r$. Obviously, the points $(kb^i, k\varphi(b^i))$ belong to conv epi$\varphi$, for all $k \geq 0$, and hence,

$$\left(\sum_{i=1}^{r} \alpha_i b^i,\ \sum_{i=1}^{r} \alpha_i \varphi(b^i)\right) \in \text{conv epi}\,\varphi. \tag{3}$$

Since there is an integer number $k^* \geq 1$ such that $M(k^* b^i)$ has only integer vertices, we have

$$\varphi^{o}(k^* b^i) = \varphi(k^* b^i) = k^* \varphi(b^i),\ i = 1,\dots,r.$$

Therefore,

$$k^*\mu \geq \varphi^{o}(k^* b) = \varphi^{o}\left(\sum_{i=1}^{r} \alpha_i k^* b^i\right) \geq \sum_{i=1}^{r} \alpha_i \varphi^{o}(k^* b^i) =$$

$$= \sum_{i=1}^{r} \alpha_i k^* \varphi(b^i),$$

from which

$$\mu \geq \sum_{i=1}^{r} \alpha_i \varphi(b^i)$$

follows. Taking the definition of epi$\varphi$ into account, one concludes from (3) that $(b,\mu) \in$ conv epi $\varphi$, which shows (2). #

Theorem 3.
Let $\varphi(0) = 0$ and let $b^* \in B$ be fixed. Then there is a number $\vartheta = \vartheta(b^*) > 0$ such that

$$\varphi(b) \geq \varphi(b^*) - \vartheta \| b - b^* \| \ \forall b \in B.$$

Proof.
For the constraint sets $G(b)$ we underly a representation

$$G(b) = K(b) + V_s,\ b \in B,$$

with an u.s.c. compact-value $K$, which exists by Theorem 5.2. (specified in Theorem 5.6.). Because of $\varphi(0) = 0$ and $(c,d)^T v \geq 0$, for all $v \in V_s$, we have

$$\psi(b) \cap K(b) \neq \emptyset \ \forall b \in B.$$

Let $\{z^1,\dots,z^r\}$ be the set of all different s-dimensional integer vectors which are possible if $(z,y)\in K(b^*)$. We choose an $\varepsilon>0$, sufficiently small, such that

$$(z,y)\in U_\varepsilon K(b^*) \Longrightarrow z\in\{z^1,\dots,z^r\}$$

holds. By the upper semicontinuity of K there is a $\delta=\delta(b^*)>0$ with

$$K(b)\subset U_\varepsilon K(b^*) \quad \forall\, b\in S_\delta$$

where

$$S_\delta=\{b\in R^m \,/\, b\in B,\ \|b-b^*\|\le\delta\}.$$

Then one has

$$b\in S_\delta\ ,\ (z,y)\in K(b) \Longrightarrow z\in\{z^1,\dots,z^r\}.$$

Using the definitions

$$\tilde\varphi(\lambda) \underset{Df}{=} \inf\{d^Ty \,/\, Qy=\lambda,\ y\geqq 0\},$$

$$I(b) \underset{Df}{=} \{i\in\{1,\dots,r\} \,/\, \exists\,(z^i,y)\in K(b)\}$$

one can write

$$\varphi(b) = \inf_{i\in I(b)}\{c^Tz^i+\tilde\varphi(b-Az^i)\} \quad \forall\, b\in S_\delta$$

where $I(b^*)=\{1,\dots,r\}$. By Theorem 1., there is a constant number $\gamma_1>0$ such that

$$|\tilde\varphi(\lambda)-\tilde\varphi(b^*-Az^i)|\le\gamma_1\|\lambda-b^*+Az^i\|,$$

for all $\lambda$ for which there exists a $y\geqq 0$ with $Qy=\lambda$ and all $i\in\{1,\dots,r\}$. From the obvious inequality

$$\tilde\varphi(\lambda)\geqq\tilde\varphi(b^*-Az^i)-|\tilde\varphi(\lambda)-\tilde\varphi(b^*-Az^i)|$$

one obtains the estimation

$$\tilde\varphi(\lambda)\geqq\tilde\varphi(b^*-Az^i)-\gamma_1\|\lambda-b^*+Az^i\|, \tag{4}$$

for all $\lambda\in\operatorname{dom}\tilde\varphi$ and $i\in\{1,\dots,r\}$.

If now $b\in S_\delta$, then there is an index $i^o\in I(b)$ such that

$$\varphi(b)=c^Tz^{i^o}+\tilde\varphi(b-Az^{i^o}),$$

which yields, by (4) and $i^o\in I(b^*)$, the estimate

$$\begin{aligned}\varphi(b)&\geqq c^Tz^{i^o}+\tilde\varphi(b^*-Az^{i^o})-\gamma_1\|(b-Az^{i^o})-(b^*-Az^{i^o})\|=\\&=c^Tz^{i^o}+\tilde\varphi(b^*-Az^{i^o})-\gamma_1\|b-b^*\|\geqq\\&\geqq\varphi(b^*)-\gamma_1\|b-b^*\|,\end{aligned}$$

for all $b\in S_\delta$.

On the other hand, if $b\notin S_\delta$, then one obtains

$$\varphi(b) \geq \varphi(b^*) - |\varphi(b) - \varphi(b^*)| \geq$$
$$\geq \varphi(b^*) - \gamma \| b-b^* \| - \eta \geq$$
$$\geq \varphi(b^*) - (\gamma + \frac{\eta}{\delta}) \| b-b^* \|$$

where by Theorem 1. (ii),

$$|\varphi(b) - \varphi(b^*)| \leq \gamma \| b-b^* \| + \eta \quad \forall b, b^* \in B$$

holds.

The proof is complete if one takes

$$\vartheta = \vartheta(b^*) = \max \{\gamma_1, \gamma + \frac{\eta}{\delta}\} . \quad \#$$

Theorem 3. will be utilized in Chapter 9. for establishing an exact penalty function related with problems P(b).

## 9. On Relations between Parametric Optimization, Solution Concepts and Subadditive Duality for Integer Optimization

In this chapter we expose such applications of parametric optimization to integer programs as they are well-known in "continuous" optimization. According to the development of the parametric analysis in the case of integer optimization these applications concern integer linear problems. When we selected the different applications for this purpose, in particular we had in mind to present, within the bounds of possibility given by the results and techniques of the previous chapters, a short and unified treatment for the most known relations.

### 9.1. Penalty Functions for (Mixed-) Integer Linear Optimization

The behaviour of penalty methods essentially depends on the involved penalty functions. In integer programming a particular difficulty appears in finding an appropriate penalty term. Using the results of Chapter 8. in this section we show that there exist exact penalty functions for pure- and mixed-integer linear optimization problems with a structure allowing the design of solution methods.
Let P(b) be the parametric problem treated in Chapter 8., and let the same assumptions be imposed as there, further, let $b^* \in B$ be a fixed right-hand side of the linear constraints. We associate with $P(b^*)$ the mixed-integer problem

$$P^* \quad \inf\{c^Tz + d^Ty + r\| b^*-Az-Qy\| / (z,y) \geq 0,\ z \text{ integer}\}.$$

By $\varphi^*$ and $\psi^*$ we denote the extremal value and the optimal set, respectively.

Theorem 1.
Let $\varphi(0) = 0$ and $b^* \in B$. If r is a sufficiently large number, then

$$\psi(b^*) = \psi^* \text{ and } \varphi(b^*) = \varphi^*.$$

Proof.
From Theorem 8.3. we conclude that

$$\varphi(b) \geq \varphi(b^*) - \vartheta\|b-b^*\| \quad \forall\, b \in R^m$$

where $\varphi(b) = +\infty$ if $b \notin B$. For $r > \vartheta$ and $(z,y) \geq 0$, z integer with $Az + Qy \neq b$, we obtain

$$c^Tz + d^Ty + r\|b^* - Az - Qy\| \geq \varphi(Az+Qy) + r\|b^* - Az - Qy\| >$$
$$> \varphi(Az+Qy) + \vartheta\|b^*-Az-Qy\| \geq \varphi(b^*),$$

from which one immediately concludes the proposition. #

We consider now the pure-integer problem

$$P(b) \quad \inf\{c^T z \ / \ Az = b,\ z \geq 0,\ z \text{ integer}\}$$

for a fixed $b \in B$. Let

$$L(z,u,r) \underset{Df}{=} (c-u^T A)^T z + r \, \|Az-b\|^2 + u^T b.$$

Theorem 2.

Let $\varphi(0) = 0$. For every $u \in R^m$ there exists a number $r(u) \geq 0$ not depending on b such that

$$L(z^*,u,r) = \min\{L(z,u,r) \ / \ z \in Z^n,\ z \geq 0\} \Longleftrightarrow z^* \in \psi(b),$$

for all $r \geq r(u)$. Moreover,

$$L(z^*,u,r) = \varphi(b).$$

Proof.

On account of $B \subset Z^m$, Theorem 8.1. yields the existence of a constant number $\varrho > 0$ such that

$$|\varphi(b) - \varphi(b')| \leq \varrho \|b-b'\| \quad \forall b,b' \in B.$$

Further, $\|b-b'\| \leq \|b-b'\|^2$ for all $b,b' \in B$.

For an arbitrarily fixed $u \in R^m$ we put

$$r(u) \underset{Df}{=} 1 + \varrho + \|u\|,$$

Now, let $z \geq 0$ integer and $Az \neq b$, and let $r \geq r(u)$. Then

$$\begin{aligned} L(z,u,r) &= c^T z + u^T(b-Az) + r \, \|Az-b\|^2 \geq \\ &\geq \varphi(Az) + u^T(b-Az) + r \, \|Az-b\| \geq \\ &\geq \varphi(b) - |\varphi(b) - \varphi(Az)| + u^T(b-Az) + r \, \|Az-b\| \geq \\ &\geq \varphi(b) - \varrho \|b-Az\| - \|u\| \, \|b-Az\| + r \, \|Az-b\| > \\ &> \varphi(b), \end{aligned}$$

which the propositions of the theorem are simple consequences of. #

Theorem 2. shows that $L(z,u,r)$ yields a penalty function (for the pure integer problem $P(b)$) which, obviously, possesses "nicer" properties than the one considered in Theorem 1.

We remark that lower bounds for the "sufficiently large" numbers r occurring in both the Theorems 1. and 2. can be derived using estimates of the constants taken from the Theorems 8.1. and 8.2..

## 9.2. Partitioning Procedures

Here we establish the general BENDERS') partitioning framework for integer programming in terms of parametric optimization. There are many concrete realizations, in particular, for special linear problems, which we will not treat here and refer the reader to the BIBLIOGRAPHIES on integer programming.

Let the general mixed-integer problem be given by

$$\inf\{f(z,y) \;/\; (z,y)\in M,\ z \text{ integer}\} \tag{1}$$

where $M \subset R^n$, $f : M \longrightarrow R$, $z \in R^s$ and $y \in R^{n-s}$.
For every $\lambda \in R^s$ we associate with M the set

$$M(\lambda) \underset{Df}{=} \{y \in R^{n-s} \;/\; (\lambda,y)\in M\} \tag{2}$$

and the extremal value

$$\varphi(\lambda) \underset{Df}{=} \inf\{f(\lambda,y) \;/\; y\in M(\lambda)\}. \tag{3}$$

Furthermore, we set

$$\Lambda \underset{Df}{=} \{\lambda \in R^s \;/\; M(\lambda) \neq \emptyset\}. \tag{4}$$

By (3) a value function $\varphi : \Lambda \longrightarrow \bar{R}$ is defined. Using the definitions above the pure integer problem

$$\inf\{\varphi(\lambda) \;/ \lambda \in \Lambda,\ \lambda \text{ integer}\} \tag{5}$$

is assigned to the mixed integer problem (1), and there is a similar relation between (1) and (5) as it was pointed out in Remark 7.1.

## The Partitioning Principle

The pair $(z^0,y^0)$ is an optimal point of the problem (3) if and only if $z^0$ is an optimal point of the pure integer problem (5), and $y^0$ is optimal for the problem

$$\inf\{f(z^0,y) \;/\; y\in M(z^0)\}. \tag{6}$$

The partitioning principle permits to design an algorithmic approach to solve problem (1) by following the idea: First one solves the pure-integer problem (5), which requires to find the set $\Lambda$ and the value function $\varphi$ in advance. If $z^0$ is an optimal point of problem (5), then one still has to solve the "continuous" optimization problem (6). Obviously, the realization and efficiency of such an approach depend on the properties of the original problem (1) and the auxiliary problems (5) and (6).
Since, for practically relevant cases, it is impossible to determine the function $\varphi$ and the set $\Lambda$ (not even in the linear case), the only thing one can expect is to follow the above idea in an approximative manner, in particular, in such a way that only an approximative determination of $\varphi$ and $\Lambda$ is necessary. In order to be successful in such a determination it is desirable to have "nice" properties of the function $\varphi$ and the set $\Lambda$ , i.e., for example, that $\varphi$ is a convex or concave function and $\Lambda$ is a convex set.

For instance, for a problem (1) with a convex objective function and a convex polyhedron M as the constraint set it is easy to show that the set $\Lambda$ is a convex polyhedron and, $\varphi$ is a convex function (if the infimum of f is finite on M).

We discuss a possible partitioning procedure for the simplest case where the objective function f is linear, too.

Let us consider

$$\inf\{c^T z + d^T y \;/\; (z,y)\in M,\; z\in R^s \text{ integer}\}$$
$$\text{where } M \underset{Df}{=} \{(z,y)\in R^n \;/\; Az + Qy \leq b,\; (z,y) \geq 0\}\;. \tag{7}$$

In order to characterize the set $\Lambda$ from (4) and the value function $\varphi$ from (3) we define the convex polyhedron

$$W \underset{Df}{=} \{u\in R^m \;/\; Q^T u \geq d,\; u \geq 0\}. \tag{8}$$

Let $U = \{u^1,\dots,u^p\}$ denote the set of all vertices of W and $H = \{h^1,\dots,h^q\}$ a generating system of the recession cone $0^+W$.

Then, from linear parametric programming, one knows (see e.g. NOŽIČKA et al.(1974)):

1° $\Lambda = \{\lambda\in R^s_+ \;/\; (b-A\lambda)^T h^i \geq 0,\; h^i\in H\}$
(if $H=\{0\}$, then $\Lambda = R^s_+$);

2° If there is a $\lambda^0\in\Lambda$ such that $\varphi(\lambda^0)$ is finite, then

$$\varphi(\lambda) = c^T\lambda + \max\{(b-A\lambda)^T u^k \;/\; u^k\in U\}\;,\; \lambda\in\Lambda.$$

If $W = \emptyset$, then (5) has no solution. Therefore we suppose that $W \neq \emptyset$.

Let

$$U'\subset U,\; U'\neq\emptyset \text{ and } H'\subset H$$

be arbitrarily chosen subsets.

We define

$$\Lambda(H') \underset{Df}{=} \{\lambda\in R^s_+ \;/\; (b-A\lambda)^T h^i \geq 0,\; h^i\in H'\},$$
$$\Lambda(\emptyset) \underset{Df}{=} R^s_+,$$
$$\varphi_{U'}(\lambda) \underset{Df}{=} c^T\lambda + \max\{(b-A\lambda)^T u^k \;/\; u^k\in U'\}.$$

Because of 1° and 2° we have

$$\Lambda\subset\Lambda(H'),$$
$$\varphi_{U'}(\lambda) \leq \varphi(\lambda),\; \lambda\in R^s.$$

Using the introduced definitions we consider the problem

$$P(H',U')\quad \inf\{\varphi_{U'}(\lambda) \;/\; \lambda\in\Lambda(H'),\; \lambda \text{ integer}\}$$

instead of the problem (5) corresponding to (7).

Let $\lambda^*$ be an optimal solution of P(H',U'). Then $\lambda^*$ is also an optimal solution of (5) if the optimality condition

$$\varphi_{U'}(\lambda^*) = c^T\lambda^* + \max\{(b-A\lambda^*)^T u \;/\; u \in W\} \;(= \varphi(\lambda^*)) \tag{9}$$

is satisfied. Otherwise, we extend the set U' or the set H' by solving the linear program

$$\max\{(b-A\lambda^*)^T u \;/\; u \in W\}. \tag{10}$$

If $u^*$ is an optimal solution of (10), then $\lambda^*$ belongs to $\Lambda$ and we have the inequality

$$\varphi_{U'}(\lambda^*) < c^T\lambda^* + (b-A\lambda^*)^T u^* (= \varphi(\lambda^*)),$$

therefore, for a better approximation, we put

$$H' := H' \text{ and } U' := U' \cup \{u^*\}.$$

If (10) has no optimal solution, then there is an unbounded edge

$$\{u \in R^m \;/\; u = u^* + th^*,\; t \geq 0\}$$

(where $u^* \in W$ is a vertex and $h^*$ is a direction of this edge)

such that

$$(b-A\lambda^*)^T h^* > 0$$

and, we redefine

$$H' := H' \cup \{h^*\},$$

and, in view of arriving at a better approximation of $\varphi(\lambda)$ one can take

$$U' := U' \cup \{u^*\}.$$

If the pure integer approximating problem P(H',U') has no optimal solution, then U' and H' have to be extended by adding vertices and edge vectors, respectively (corresponding to unbounded edges of W) not considered up to now. Obviously, only a finite number of such iterations are possible and, if one arrives at $U' \equiv U$, $H' \equiv H$ and P(H,U) is unsolvable, then the original problem (7) has no optimal solution.

The efficiency of such an iterative approach depends essentially on the initial sets U' and H'. In general, it is hard to decide which subsets of U and H should be used. A possible initialization could be obtained by solving the linear programming relaxation of (7). If this linear program is unsolvable in spite of the fact that the matrices A and Q are rational, then the corresponding integer problem (7) has also no optimal solution. If now $(z^*, y^*)$ is an optimal solution of the linear programming relaxation and $u^*$ is an optimal vertex of (10), for $\lambda^* = z^*$, then one can choose $u^*$ and its adjacent vertices (or some of them) as U' and the corresponding vectors in the direction of unbounded edges (if there are some) as H'. The problems P(H',U') appearing at every iteration step can be solved by a modified algorithm which solves pure integer linear programs since P(H',U') is equivalent to

$$\inf\{v \ / \ v \geq c^T\lambda + (b-A\lambda)^T u^k,\ u^k \in U';\ v \in R,$$
$$(b-A\lambda)^T h^i \geq 0,\ h^i \in H',\ \lambda \in R^s_+ \text{ integer}\}$$

Thereby, one can utilize all the information obtained in the previous step because at most two new constraints occur in the current problem $P(H',U')$.

## 9.3. Subadditive Duality

This section is motivated by the fact that the value function of the parametric problem P(b) of Chapter 8. is an optimal solution of the subadditive dual program for (mixed-) integer programs, which first was implicitly used by BURDET/JOHNSON (1974) and explicitly studied by JEROSLOW (1978) in order to characterize valid cuts and minimal inequalities.

Our considerations start with the parametric problem P(b) from Chapter 8., i.e.

$$P(b)\ \inf\{c^T z + d^T y \ / \ x = (z,y) \in G(b)\},\ b \in R^m,$$

where

$$G(b) \underset{Df}{=} \{x \in R^n \ / \ x = (z,y),\ Az + Qy = b,\ z,y \geq 0,\ z \in R^s,\ \text{integer}\}$$

under the general assumption that the matrix

$$D \underset{Df}{=} (A,Q) \text{ is rational.}$$

As in Chapter 8. the general assumption is imposed in order to obtain a unified treatment. We note that some results hold under weaker assumptions.

If $S \subset R^n$ is a monoid, then a function $f : S \rightarrow R \cup \{-\infty\}$ is said to be subadditive when

$$f(x+y) \leq f(x) + f(y) \qquad \forall x,y \in S.$$

If g is a function from S into $R \cup \{-\infty\}$ then the limit

$$g'(x) \underset{Df}{=} \limsup_{t \to 0_+} \frac{g(tx)}{t}$$

is called the upper directional derivative of g at zero, where $tx \in S$, $0 \leq t \leq \varepsilon$, for some $\varepsilon > 0$, is assumed (see e.g. JOHNSON 1974).

The domain B of G is a monoid (see Remark 5.1). The following Theorem 3. expresses the role of the value function

$$\varphi(b) \underset{Df}{=} \inf\{c^T z + d^T y \ / \ (z,y) \in G(b)\}$$

of P(b) for the definition of a subaddition dual program:

Theorem 3.

If $\varphi(0) = 0$, then

(i) the value function $\varphi$: $B \longrightarrow R$ is subadditive;

(ii) $\varphi(A^i) \leq c_i$, $i = 1,\ldots,s$,

$\varphi'(Q^j) \leq d_j$, $j = 1,\ldots,n-s$ ;

(iii) any subadditive function $g : B \longrightarrow R \cup \{-\infty\}$ having the properties of (ii) and satisfying $g(0) = 0$ fulfils the inequality

$$\varphi(b) \geq g(b) \ \forall b \in B.$$

Proof. (i) From $x \in \psi(b)$, $x' \in \psi(b')$ and $(x+x') \in G(b+b')$ it follows that $\varphi(b) + \varphi(b') = (c,d)^T x + (c,d)^T x' \geq \varphi(b+b')$.

(ii) $\varphi(A^i) \leq c_i$ is trivial. Obviously, one has $\varphi(tQ^j) \leq t\, d_j$, from which $\frac{\varphi(tQ^j)}{t} \leq d_j$ and, consequently, $\varphi'(Q^j) \leq d_j$ follow.

(iii) One easily sees that if a subadditive function g satisfies $g(0) = 0$, then

$$g(tb) \leq tg'(b) \quad \forall t \geq 0.$$

If $b \in B$ is arbitrarily fixed and $x^0 = (z^0,y^0) \in \psi(b)$, then

$$g(b) = g(\sum_{i=1}^{s} A^i z_i^0 + \sum_{j=1}^{n-s} Q^j y_j^0) \leq \sum_{i=1}^{s} g(A^i) z_i^0 + \sum_{j=1}^{n-s} g(Q^j y_j^0) \leq$$

$$\leq \sum_{i=1}^{s} c_i z_i^0 + \sum_{j=1}^{n-s} g'(Q^j) y_j^0 \leq \sum_{i=1}^{s} c_i z_i^0 + \sum_{j=1}^{n-s} d_j y_j^0 = \tag{11}$$

$$= \varphi(b). \quad \#$$

For every fixed $b \in R^m$ the following problem

SD(b) $\sup \{g(b) \ / \ b \in F\}$

where

$$F \underset{Df}{=} \{g \ / \ g: B \longrightarrow R \cup \{-\infty\} \text{ is a subadditive function; } g(A^i) \leq c_i, \ i = 1,\ldots s; \ g'(Q^j) \leq d_j, \ j = 1,\ldots,n-s; \ g(0) = 0\} \tag{12}$$

is called the <u>subadditive dual program</u> to P(b). Indeed, that SD(b) is a dual program to P(b) becomes clear by the following Theorem 4., which is a strong duality theorem:

Theorem 4.

The following statements hold, for every fixed $b \in R^m$ :

(i) the two problems P(b) and SD(b) have optimal solutions if and only if both feasible sets G(b) and F are non-empty;

(ii) if $x^0 = (z^0,y^0)$ and $g^0$ are optimal solutions of P(b) and SD(b), respectively, then

1° $c^T z^0 + d^T y^0 = \varphi(b) = g^0(b)$,

2° $\sum_{i=1}^{s} z_i^0 (c_i - \varphi(A^i)) + \sum_{j=1}^{n-s} y_j^0 (d_j - \varphi'(Q^j)) = 0$;

(iii) if $x^o = (z^o, y^o)$ is an optimal point of P(b), then

$$\varphi(A^i) = \varphi(b) - \varphi(b-A^i) \text{ if } z_i^o > 0,$$
$$\varphi(\varepsilon Q^j) = \varphi(b) - \varphi(b-\varepsilon Q^j) \ \forall \varepsilon : 0 \leq \varepsilon \leq y_j^o \text{ if } y_j^o > 0.$$

Proof. (i) $\Psi(b) \neq \emptyset$ implies $\varphi(0) = 0$ and, by Theorem 3., it follows that $\varphi$ is optimal in SD(b). If $G(b) \neq \emptyset$ and $F \neq \emptyset$, then we have to distinguish between the two cases (a) $\varphi(b)$ is finite and, (b) $\varphi(b) = -\infty$. If the case (a) arises, then everything is clear, by Theorem 3. In the case (b) one has $\varphi(b) = -\infty$, for all $b \in B$. For any $g \in F$, by Theorem 3., $0 = g(0) \leq \varphi(0)$, but this is impossible.

(ii) 1° is clear. By (11), one has

$$\varphi(b) \leq \sum_{i=1}^{s} \varphi(A^i) z_i^o + \sum_{j=1}^{n-s} \varphi'(Q_j) y_j \leq \sum_{i=1}^{s} c_i z_i^o + \sum_{j=1}^{n-s} d_j y_j^o = \varphi(b).$$

By $(z_i^o, y_j^o) \geq 0$, it follows from Theorem 3. (ii) that

$$z_i^o = 0 \quad \text{if } \varphi(A^i) < c_i,$$
$$y_j^o = 0 \quad \text{if } \varphi'(Q^j) < d_j;$$

hence, 2° is true.

(iii) Let $x^o = (z^o, y^o) \in \Psi(b)$. We define the points $x^i(\varepsilon) \underset{Df}{=} x^o - \varepsilon e^i$ where $e^i \in R^n$ denotes the i-th unit vector and $\varepsilon \geq 0$.

Obviously, $e^i \in G(A^i)$ and, $x^i(1) \in G(b-A^i)$ if $z_i^o > 0$ (i.e. actually $z_i^o \geq 1$). Hence,

$$\begin{aligned} \varphi(b) = (c,d)^T x^o &= (c,d)^T x^i(1) + (c,d)^T e^i \geq \\ &\geq \varphi(b-A^i) + \varphi(A^i). \end{aligned} \tag{13}$$

If $y_j^o > 0$ and $0 \leq \varepsilon \leq y_j^o$, then we have $x^o = x^{s+j}(\varepsilon) + \varepsilon e^{s+j}$, $\varepsilon e^{s+j} \in G(\varepsilon Q^j)$ and $x^{s+j}(\varepsilon) \in G(b-\varepsilon Q^j)$, which implies

$$\begin{aligned} \varphi(b) &= (c,d)^T x^{s+j}(\varepsilon) + \varepsilon (c,d)^T e^{s+j} \geq \\ &\geq \varphi(b-\varepsilon Q^j) + \varphi(\varepsilon Q^j). \end{aligned} \tag{14}$$

By the subadditivity of $\varphi$ also the converse inequalities to (13) and (14) are true, from which the proposition follows. #

Theorem 4. shows that one can state similar propositions as in the case of a pair of dual linear programs for the pair P(b) and SD(b), for all fixed b. However, in contrast to the linear pair, the dual program here has subadditive functions as variables and the feasible set F of SD(b) does not exihibit a tractable structure. Thus, it is desirable to restrict the dual variables to a smaller class of functions whose structure allows better applications and at least guarantees weak duality (i.e. equality of the extremal values). The additional requirement on the dual functions to be convex on $\Lambda$ = conv B (=domM) does not lead to the desired goal.

If

$$F^* \underset{Df}{=} \{g: \Lambda \longrightarrow \mathbb{R} \cup \{-\infty\} \ / \ g \text{ subadditive, convex and } g \in F\}$$

and $g \in F^*$, then one obtains

$$g(\lambda b) = \lambda g(b) \ \forall \lambda \geq 0 \ \forall b \in \Lambda .$$

Therefore, one easily verifies that

$$\varphi^o(b) \underset{Df}{=} \inf\{(c,d)^T x \ / \ (A,Q)x = b, \ x \geq 0\} \geq g(b) \forall b \in \Lambda.$$

Now, since $\varphi^o \in F^*$ (if $\varphi^o(0) = 0$), one directly could use the dual to $P^o(b)$

$$D^o(b) \ \sup\{b^T u \ / \ \begin{pmatrix} A^T \\ Q^T \end{pmatrix} u \leq \begin{pmatrix} c \\ d \end{pmatrix}\}$$

instead of

$$SD^*(b) \ \sup\{g(b) \ / \ g \in F^*\}$$

to be the dual program to P(b).

But, this would mean that the integer requirements would not play any role and, obviously, even weak duality would fail . However, from Theorem 3.(iii), we know that the value function $\varphi$ of P(b) is optimal for the subadditive dual SD(b), for all $b \in B$ (if $\varphi(0) = 0$ is assumed). Hence a deeper characterization of the value function could lead to a reasonable restriction of the feasible set F corresponding to the dual problem SD(b). BLAIR/JEROSLOW (1982) could show for the pure integer case that the value function $\varphi$ of P(b) is a so-called Gomory function and, since every Gomory function may be represented by the maximum of a finite number of so-called Chvátal functions, one can restrict the dual variables in this case to Chvátal functions. We will give here a short outline of this approach, for the proofs the reader is referred to the original paper of BLAIR/JEROSLOW (1982).

The class of m-dimensional Chvátal functions is the smallest class $\mathcal{C}$ of functions obtained in an iterative way: Starting with all linear functions $f: Q^m \longrightarrow Q$ (here Q denotes the field of all rational numbers) and finitely often repeating the operations of taking nonnegative rational linear combinations, i.e.

$$(\alpha f+\beta g)(b) \underset{Df}{=} \alpha f(b) + \beta g(b), \alpha, \beta \in Q_+, \ f, g \in \mathcal{C},$$

and rounding up to the nearest integer, i.e.

$$\lceil f \rceil (b) \underset{Df}{=} \lceil f(b) \rceil, \ f \in \mathcal{C} .$$

In essence, the operations defining the class $\mathcal{C}$ of Chvátal functions are those which are well-known from cutting-plane theory (e.g. GOMORY (1963) and CHVÁTAL (1973)), which generate new valid cuts from a valid inequality. Moreover, the following shows that Chvátal functions are a discrete analogue to linear functions, and it is the merit of BLAIR/JEROSLOW to have found out this fact.

The class $\mathcal{G}$ of <u>m-dimensional Gomory functions</u> contains the class $\mathcal{C}$ and all functions which are obtained by finitely often repeating the taking of the maximum of functions of $\mathcal{C}$ and functions already obtained by these operations, i.e.

$$f, g \in \mathcal{G} \Longrightarrow \max\{f, g\} \in \mathcal{G}$$

where $\max\{f,g\}(b) \underset{Df}{=} \max\{f(b), g(b)\}$.

It is easy to verify that a function $f \in \mathcal{G}$ is subadditive on the monoid $Q^m$ and that $f \in \mathcal{G}$ is representable by the maximum of a finite number of functions from $\mathcal{C}$.

BLAIR/JEROSLOW showed that Gomory functions provide value functions of pure integer problems P(b).

Using ideas from cutting-plane theory they showed in a constructive way that the following theorem is true.

<u>Theorem 5.</u>

Let P(b) be a pure integer problem. Then there is a Gomory function $f : Q^m \longrightarrow Q$ such that $b \in B$ if and only if $f(b) \leq 0$.

Based on the **preceding** Theorem 5. they proved that the value function of P(b) is a Gomory function, i.e. a maximum of a finite number of Chvátal functions.

<u>Theorem 6.</u>

Let P(b) be a pure integer problem and $c \in Q^n$ be fixed. If $\varphi(0) = 0$, then there is a Gomory function $g : Q^m \longrightarrow Q$ with

$$g(b) = \varphi(b), \forall b \in B.$$

Now, by the Theorems 6. and 3. and the above discussion, it becomes clear that the following corollary holds:

<u>Corollary 1.</u>

If the pure integer problem P(b) is solvable and $c \in Q^n$, then the subadditive dual SD(b) has an optimal solution which is a Chvátal function.

On account of the Theorems 5.,6. and Corollary 1. one easily verifies that Chvátal functions are discrete **analogues** to linear functions with respect to linear and pure integer linear programming (under right-hand side perturbations). While the domain B may be described by a finite system of linear inequalities if no integer requirements are imposed, the corresponding domain can be expressed by a finite number of inequalities with Chvátal functions if the pure integer case is considered (Theorem 5.).

The value function $\varphi^0$ of $P^0(b)$ is the maximum of finitely many linear functions and, $\varphi$ is the maximum of a finite number of Chvátal functions (Theorem 6.) in the pure integer case.

Corollary 1. shows a corresponding relation between the dual variables.

# BIBLIOGRAPHY

Allende, S. (1980). Globale Stabilitätsuntersuchungen rein- und gemischtganzzahliger parametrischer Optimierungsprobleme. Diss.(A), Humboldt -Universität Berlin.

Andronov, V.G., E.G. Belousov, and I.V. Shironin (1982). On the solvability of polynomial programming problems. (Russian) Izv. Akad. Nauk SSSR, Ser. Tekhn. Kibernet., no. 4, 194-197.

Andronov, V.G., E.G. Belousov (1985). On the stability of polynomial optimization problems with linear constraints. (Russian) Izv. Akad. Nauk SSSR, Ser. Tekhn. Kibernet., no. 3, 68-77.

Bachem, A., R. Schrader (1978). A note on a theorem of Jeroslow. Report No. 7897-OR Institut für Operations Research, Universität Bonn.

Bank, B. (1977). Qualitative Stabilitätsuntersuchungen rein- und gemischtganzzahliger linearer parametrischer Optimierungsprobleme. Diss. (B), Humboldt-Universität Berlin.

Bank, B., E.G. Belousov, R. Mandel, J.N. Ceremnych, and V.M. Shironin (1986). Mathematical programming: Questions of solvability and stability. (Russian) Edition Moscow University.

Bank, B., J. Guddat, D. Klatte, B. Kummer, and K. Tammer (1982). Nonlinear Parametric Optimization. Akademie-Verlag, Berlin.

Bank, B., R. Hansel (1984). Stability of mixed-integer quadratic programming problems. Mathematical Programming Study 21, 1-17, North-Holland.

Bank, B., R. Mandel (1986). Nonlinear parametric integer programming. In: Guddat, J. et al. (eds.), Proceedings of the International Conference on 'Parametric Optimization and Related Topics'. Plaue/DDR, Okt. 1985, Akademie-Verlag Berlin.

Bank, B., R. Mandel, and K. Tammer (1982). Parametrische Optimierung und Aufteilungsverfahren. In: Lommatzsch, K. (ed.). Anwendungen der linearen parametrischen Optimierung. Akademie-Verlag, Berlin.

Belousov, E.G. (1977). Introduction to convex analysis and integer programming. (Russian) Edition Moscow University.

Belousov, E.G. (1986). Stability of polynomial optimization problems. In: Guddat, J. et al. (eds.). Proceedings of the International Conference on 'Parametric Optimization and Related Topics'. Plaue/DDR, Okt. 1985. Akademie-Verlag, Berlin.

Benders, J.F. (1962). Partitioning procedures for solving mixed-variables programming problems. Numerische Mathematik 4,3, 238-252.

Bibliography I. (1976). Integer programming and related areas. Ed. C. Kastning, Lecture Notes in Economics and Mathematical Systems, 128. Springer-Verlag, Berlin-Heidelberg-New York.

Bibliography II. (1978). Integer programming and related areas. Ed. D. Hausmann, Lecture Notes in Economics and Mathematical Systems, 160. Springer-Verlag, Berlin-Heidelberg-New York.

Bibliography III. (1984). Integer programming and related areas. Ed. P. von Randow. Lecture Notes in Economics and Mathematical Systems, 197. Springer-Verlag, Berlin-Heidelberg-New York.

Blair, C.E., R.G. Jeroslow (1977). The value function of a mixed integer program: I. Discrete Mathematics, 19, 121-138.

Blair, C.E., R.G. Jeroslow (1979). The value function of a mixed integer programm: II. Discrete Mathematics, 25, 7-19.

Blair, C.E., R.G. Jeroslow (1981). An exact penalty method for mixed-integer programs. Mathematics of operations research, 6, No. 1, 14-18.

Blair, C.E., R.G. Jeroslow (1982). The value function of an integer program. Mathematical Programming 23, 237-237.

Blum, E., W. Oettli (1972). Direct proof of the existence theorem for quadratic programming. Operations Research, 20, 165-167.

Borevich, Z.J. (1966). Number Theory. Academic Press,New York and London.

Borosh, I., L.B. Treybig (1976). Bounds on positive integral solutions of linear Diophantine equations. Proc. Amer. Math.Soc. 55, 299-304.

Bowman, V.J., C.-A. Burdet (1974). On the general solution to systems of mixed-integer linear equations. SIAM J. Appl. Math. 26, No. 1.

Brosowski, B. (1982). Parametric Semi-Infinite Optimization. Verlag Peter D. Lang, Frankfurt/M.-Bern.

Burdet, C.-A., E.L. Johnson (1974). A subadditive approach to the group problem of integer programming. Mathematical Programming Study 2, 51-71.

Burkard, R.E. (1972). Methoden der Ganzzahligen Optimierung. Springer-Verlag, Wien-New York.

Cassels, J. W.S. (1957). An Introduction to Diophantine Approximation. Cambridge Univ. press, Cambridge.

Charnes, A., F. Granot (1975). Existence and representation of diophantine and mixed diophantine solutions to linear equations and inequalities. Discrete Mathematics, 11, 233-248.

Chvátal, V. (1973). Edmonds Polytopes and a Hierarchy of Combinatorial Problems. Discrete Mathematics 4, 305-337.

Collins, G.E. (1975). Quantifier elimination for real closed fields by cylindrical algebraic decomposition. In: Lecture Notes in Computer Science 33, 134-183.

Dinkelbach, W. (1969). Sensitivitätsanalysen und parametrische Programmierung. Springer-Verlag, Berlin-Heidelberg-New York.

Dontchev, A.L. (1983). Perturbations, Approximations and Sensitivity Analysis of Optimal Control Systems. Lecture Notes Control Inform. Sci. Springer-Verlag, Berlin-Heidelberg-New York.

Eaves, C.B. (1971). On quadratic programming. Management Science. Vol. 17, 698-711.

Fiacco, A.V. (1983). Introduction to Sensitivity and Stability Analysis in Nonlinear Programming. Academic Press, New York.

Fiorot, J.Ch. (1972). Generation of all integer points for given sets of linear inequalities. Math. Programming, 3, 276-295.

Fisher, M.L., J.F. Shapiro (1974). Constructive duality in integer programming. SIAM J. Appl. Math. 27, 31-52.

Frank, C.R. (1967). Parametric programming in integers. Operations-Res.-Verfahren. Bd III, 167-180. Meisenheim: A. Hain.

Gal, T. (1979). Postoptimal Analysis Parametric Programming and Related Topics. McGraw-Hill, New York.

Gal, T. (1973). Betriebliche Entscheidungsprobleme, Sensitivitätsanalyse und parametrische Optimierung. Berlin, New-York, Walter de Gruyter.

Geoffrion, A.M., (1971). Duality in nonlinear programming: A simplified applications-oriented development. SIAM Review 13, 1-37.

Geoffrion, A.M. (1974). Lagrangean relaxation for integer programming. Mathematical Programming Study 2, 82-114.

Geoffrion, A.M., R. Nauss (1977). Parametric and postoptimality analysis in integer linear programming. Management Science, Vol. 23, No. 5, 453-466.

Gomory, R.E. (1963). An algorithm for integer solutions to linear programs. In: Graves and Wolfe (eds.), Recent Advances in Mathematical Programming, McGraw-Hill, New York, 269-302.

Gomory, R.E. (1965). On the Relations between Integer and Non-Integer Solutions to Linear Programs. Proc. of the National Acad. of Sciences, 53, 260-265.

Gomory, R.E. (1969). Some polyhedra related to combinatorial problems. J. Linear Alg. Appl., 2, 451-558.

Grunewald, F.J., D. Segal (1981). How to solve a quadratic equation in integers? Math. Proc. Camb. Phil. Soc., 89, 1-5.

Gurari, E.M., O.H. Ibarra (1979). An NP-complete number-theoretic problem. Journ. of the ACM, Vol. 26, No. 3, 567-581.

Hansel, R. (1980). Stabilität gemischt-ganzzahliger quadratischer Optimierungsprobleme Diss. (A), Humboldt-Universität, Berlin.

Hoffman, A.J. (1952). On approximate solutions of systems of linear inequalities. J. Res. Nat. Bur. Standards, 49, 263-265.

Holm, S., D. Klein (1978). Discrete right-hand side parametrization for linear integer programming. European Jour. of Oper. Res., 2, 50-53.

Jeroslow, R.G. (1978). Cutting plane theory: algebraic methods. Discrete mathematics, 23, 121-150.

Jeroslow, R.G. (1978a). Some basis theorems for integral monoids, Reprint No. 818. Carnegie-Mellon Univ., Pittsburgh.

Jeroslow, R.G. (1979). Minimal inequalities. Mathematical Programming 17, 1-15.

Khachiyan, L.G. (1982). Convexity and algorithmic complexity of polynomial programming problems. (Russian) Izv. Akad. Nauk SSSR, Ser. Tekhn. Kibernet., no. 6, 46-56.

Khachiyan, L.G., S.P. Tarasov (1980). Bounds and algorithmic complexity of convex Diophantine inequalities. (Russian) Dokl. Akad. Nauk SSSR, 225.

Klatte, D. (1977). Untersuchungen zur lokalen Stabilität konvexer parametrischer Optimierungsprobleme. Diss. (A), Humboldt-Universität zu Berlin.

Klatte, D. (1984). Beiträge zur Stabilitätsanalyse nichtlinearer Optimierungsprobleme. Diss. (B), Humboldt-Universität, Berlin.

Klatte, D. (1986). Lipschitz continuity of infima and optimal solutions in parametric optimization: The polyhedral case. In: Guddat, J. et al. (ed.). Proceedings of the International Conference on 'Parametric Optimization and Related Topics', Plaue/DDR, Okt. 1985, Akademie-Verlag, Berlin.

Klatte, D., B. Kummer (1987). On Stability Analysis in Nonlinear Optimization. Mathematische Forschung. Akademie-Verlag Berlin.

Koserazkaja, L.N., T.T. Lebedeva, and J.V. Sergienko (1983). Questions of stability, parametrization and postoptimality analysis of discrete optimization problems. (Russian) Kibernetika (Kiev), no. 4, 71-80.

Kummer, B. (1977). Globale Stabilität quadratischer Optimierungsprobleme. Wiss. Z. der Humboldt-Univ. Berlin, Math.-Nat. R., XXVI, 5, 565-569.

Kummer, B. (1978). Stability and weak duality in convex programming without regularity. Preprint 1978; inzwischen erschienen in: Wiss. Z. der Humboldt-Univ. Berlin, Math.-Nat. R., XXX, 5, 381-386.

Libura, M. (1976). Stability regions for optimal solutions of the integer programming problems. In: Working Paper MPD 3-76, Warsaw: System Res. Inst. Pol. Acad. Sci. 1976.

Mandel, R. (1981). Über die Existenz von Lösungen ganzzahliger Optimierungsaufgaben. Math. Operationsforschung u. Statistik, Ser. Optimization, Heft 1, 33-39.

Mandel, R. (1983). Zur Konstruktivität bei (gemischt-) ganzzahligen Optimierungsaufgaben. Math. Operationsforschung u. Statistik, Ser. Optimization, Heft 3, 343-357.

Mandel, R. (1985). Beiträge zur Theorie ganzzahliger Optimierungsaufgaben. Dissertation (B), Humboldt-Universität zu Berlin.

Marsten, R.E., T.L. Morin (1977). Parametric integer programming: The right-hand side case. In: Annals of Discrete Mathematics 1, Amsterdam, North-Holland 375-390.

Mathematical programming Study, 10 (1979). Point to Set maps and Mathematical Programming. (ed. P. Huard). North-Holland-Publ. Comp., Amsterdam.

Mathematical Programming Study, 19 (1982). Optimality and Stability in Mathematical Programming. (ed. M. Gaiguard). North-Holl.-Publ.Comp., Amsterdam.

Mathematical Programming Study, 21 (1984). Sensitivity, Stability and Parametric Analysis. (ed. A.V. Fiacco). North-Holl.-Publ. Comp., Amsterdam.

Matijasevič, Y. (1970). Enumerable sets are Diophantine. Dokl. Akad. Nauk SSSR, 191, 279-282 (Russian).

Meyer, R.R. (1974). On the existence of optimal solutions to integer and mixed-integer programming problems. Mathematical Programming, 7, 223-235.

Meyer, R.R. (1975). Integer and mixed integer programming models: General properties. J. Optim. Theory Appl., 16, 191-206.

Meyer, R.R. (1976). Mixed-integer minimization models for piecewise linear functions of a single variable. Discrete Math., 16, 163-171.

Noltemeier, H. (1970). Sensitivitätsanalyse bei diskreten linearen Optimierungsproblemen. Lecture Notes in Operation Research and Mathematical Systems, Springer-Verlag, Berlin-Heidelberg-New York.

Nožička, F., J. Guddat, H. Hollatz, and B. Bank (1974). Theorie der linearen parametrischen Optimierung. Akademie-Verlag Berlin.

Piehler, J. (1966). Ganzzahlige Optimierung. B.G. Teubner, Leipzig.

Radke, M. (1975). Sensitivity analysis in discrete optimization. Working paper 240, Western Management Science Institute, Univ. of California, Los Angeles.

Rockafellar, R.T. (1970). Convex Analysis. Princeton University Press, Princeton.

Schrijver, A. (1979). On cutting planes. Department of Mathematics Eindhoven University.

Seeländer, J. (1980). Einige Bemerkungen zur Bestimmung von Stabilitätsbereichen in der reinganzzahligen linearen Optimierung. Mathem. Operationsforschung und Statistik,Ser. Optimization,11,No. 2, 261-271.

Seeländer, J. (1980a). Über die Bestimmung von Stabilitätsbereichen bei speziellen gemischtganzzahligen linearen Optimierungsproblemen. Math. Operationsforschung und Statistik, Ser.Optimization,11,No. 3,447-454.

Shapiro, J.F. (1977). Sensitivity analysis in integer programming. In: Annals of Discrete Mathematics 1, Amsterdam, North-Holland, 467-478.

Shironin, V.M. (1980). On the distribution of integer points in unbounded convex sets in the euclidean space. (Russian) Diss. (cand. fiz. mat. nauk) Moscow WZ Akad. Nauk SSSR.

Stoer, J., C. Witzgall (1970). Convexity and Optimization in finite dimensions I. Springer-Verlag, Berlin-Heidelberg-New York.

Tarski, A. (1951). A decision method for elementary algebra and geometry. University of California Press.

Tind, J., L.A. Wolsey (1981). An elementary survey of general duality theory in mathematical programming. Math. Prog. 21, 241-261.

Van den Dries, L. (1981). Which curves over Z have points with coordinates in a discrete ordered ring? Trans. of the Amer. Math. Soc., 264, 1, 181-189.

Van de Panne, C. (1975). Methods for Linear and Quadratic Programming. Amsterdam: North-Holl. Publ. Comp.

Walkup, D., R. Wets (1969). A Lipschitzian Characterization of Convex Polyhedra. Proc. Amer. Math. Soc., 23, 167-173.

Wolsey, L.A. (1971). Group-theoretic results in mixed integer programming. Operations Research, 19, 1691-1697.

Wolsey, L.A. (1981). Integer programming duality: Price functions and sensitivity analysis. Mathematical Programming, 20, 173-195.

Wüthrich, H.R. (1976). Ein Entscheidungsverfahren für die Theorie der reell-abgeschlossenen Körper. In: Lecture Notes in Computer Science, 43, 138-162.

Band 1

**Inverse and Improperly Posed Problems in Differential Equations**

Proceedings of the Conference on Mathematical and Numerical Methods, held in Halle/Saale (GDR) from May 29 to June 2, 1979

Edited by Gottfried Anger
296 Seiten, 1979 (vergriffen)

Band 2

**Fundamentals of Computation Theory FCT '79**

Proceedings of the Conference on Algebraic, Arithmetic and Categorial Methods in Computation Theory, held in Berlin/Wendisch-Rietz (GDR) September 17–21, 1979

Edited by Lothar Budach
576 Seiten, 1979

Band 3

**Decidability and Generalized Quantifiers**

By Andreas Baudisch, Detlef Seese, Hans-Peter Tuschik, and Martin Weese
XII, 235 Seiten, 1980

Band 4

**Zur Elastizitäts- und Thermoelastizitätstheorie**

Potentialtheoretische Methoden in Anwendung auf verschiedene Modelle

Von Lothar Jentsch und Johannes Maul
157 Seiten, 1980

Band 5

**Systems Analysis and Simulation 1980**

Proceedings of the International Symposium held at Berlin (GDR) September 1–5, 1980

Edited by Achim Sydow
496 Seiten, 1980 (vergriffen)

Band 6

**Nichtlineare diskrete Optimierung**

von Eberhard Girlich und Michail M. Kowaljow
218 Seiten, 1981

Band 7

**Completeness Problems in the Structural Theory of Automata**

By Jürgen Dassow
148 Seiten, 1981

Band 8

**Darstellung von Lösungen linearer elliptischer Differentialgleichungen**

Von Günther Wildenhain
92 Seiten, 1981

Band 9

**Introduction to the Spectral Theory of Operators in Spaces with an Indefinite Metric**

By I. S. Iohvidov, M. G. Krein, and H. Langer
120 Seiten, 1982 (vergriffen)

Band 10

**Nonlinear Time-discrete Systems**

A General Approach by Nonlinear Superposition

By Michael Gössel
115 Seiten, 1982 (vergriffen)

Band 11

**Preprimal Algebras**

By Klaus Denecke
162 Seiten, 1982

Band 12

**Ergodic Theory and Related Topics**

Proceedings of the Conference held in Vitte/Hiddensee (GDR) October 19–23, 1981

Edited by Horst Michel
232 Seiten, 1982

Band 13

**Singularly Perturbed Differential Equations**

By H. Goering, A. Felgenhauer, G. Lube, H.-G. Roos, and L. Tobiska
176 Seiten, 1983

Band 14

**Leben wir in einer Volterra-Welt?**

Ein ökologischer Zugang zur angewandten Systemanalyse

Von Manfred Peschel und Werner Mende
233 Seiten, 1983

Band 15

**Markov Decision Problems with Countable State Spaces**

Optimality Criteria – Algorithms – Clustering

By Hans Michael Dietz and Volker Nollau
174 Seiten, 1983

Band 16

**Stochastische Verfahren zur Suche nach dem Minimum einer Funktion**

Von Ryszard Zieliński und Peter Neumann
133 Seiten, 1983

Band 17

**Invertibility and Asymptotics of Toeplitz Matrices**

By Albrecht Böttcher and Bernd Silbermann
200 Seiten, 1983

Band 18

**Structural Induction on Partial Algebras**

Introduction to Theory and Application of Partial Algebras – Part II

By Horst Reichel
205 pages, 1984

Band 19

**Algebraic Methods for Toeplitz-like Matrices and Operators**

By Georg Heinig and Karla Rost
212 pages, 1984

Band 20

**Frege Conference 1984**

Proceedings of the International Conference held at Schwerin (GDR), September 10–14, 1984

Edited by Gerd Wechsung

408 pages, 1984

Band 21

**Numerical Analysis of Selected Semilinear Differential Equations**

Edited by Christian Großmann, Thomas Riedrich, and Hartmut Schönheinz

203 pages, 1984

Band 22

**Singular Ordinary Differential Operators and Pseudodifferential Equations**

By Johannes Elschner

200 pages, 1985

Band 23

**Lotka-Volterra-Approach to Cooperation and Competition in Dynamic Systems**

Proceedings of the 5th Meeting of UNESCO's Working Group on Systems Theory held on the Wartburg, Eisenach (GDR) at March 5–9, 1984

Edited by Werner Ebeling and Manfred Peschel

271 pages, 1985

Band 24

**Convergence Structures 1984**

Proceedings of the Conference on Convergence held in Bechyně (Czechoslovakia), September 24–28, 1984

Edited by Josef Novák, Werner Gähler, Horst Herrlich, and Jan Mikusiński

254 pages, 1985

Band 25

**Parcella '84**

Proceedings of the II. International Workshop on Parallel Processing by Cellular Automata and Arrays held in Berlin (GDR), September 25–27, 1984

Edited by Wolfgang Händler, Tamás Legendi, and Gottfried Wolf

239 pages, 1985

Band 26

**Multiobjective and Stochastic Optimization Based on Parametric Optimization**

By Jürgen Guddat, Francisco Guerra Vasquez, Klaus Tammer, and Klaus Wendler

175 pages, 1985

Band 27/28

**Systems Analysis and Simulation 1985**
**I: Theory and Foundations**
**II: Applications**

Proceedings of the International Symposium held in Berlin (GDR), August 26–31, 1985

Edited by Achim Sydow, Manfred Thoma, and Robert Vichnevetsky

per volume 495 pages, 1985

Band 29

**Parcella '86**

Proceedings of the III. International Workshop on Parallel Processing by Cellular Automata and Arrays held in Berlin (GDR), September 9–11, 1986

Edited by Tamás Legendi, Dennis Parkinson, Roland Vollmar, and Gottfried Wolf

272 pages, 1986

Band 30

**Fuzzy Sets Applications, Methodological Approaches, and Results**

Proceedings of the International Workshop on Fuzzy Sets Applications cosponsored by IIASA and convened in cooperation with IAEA held on the Wartburg, Eisenach (GDR), at March 3–8, 1985

Edited by Steffen Bocklisch, Sergei Orlovski, Manfred Peschel, and Yasushi Nishiwaki

233 pages, 1986

Band 31

**Mathematical Methods of Specification and Synthesis of Software Systems '85**

Proceedings of the International Spring School, held in Wendisch-Rietz (GDR), April 22–26, 1985

Edited by Wolfgang Bibel and Klaus P. Jantke

245 pages, 1986

Band 32

**A Model Theoretic Oriented Approach to Partial Algebras**

Introduction to Theory and Application of Partial Algebras – Part I

By Peter Burmeister

319 pages, 1986

Band 33

**Finite-difference Methods on Irregular Networks**

A Generalized Approach to Elliptic Problems of Second Order

By Bernd Heinrich

206 pages, 1986

Band 34

**Modern Algorithms for Large Sparse Eigenvalue Problems**

By Arnd Meyer

125 pages, 1987

Band 35

**Parametric Optimization and Related Topics**

Edited by Jürgen Guddat, Hubertus Th. Jongen, Bernd Kummer, and František Nožička

407 pages, 1987

Band 36

**Discretization in Differential Equations and Enclosures**

Edited by Ernst Adams, Rainer Ansorge, Christian Großmann, and Hans-Görg Roos

250 pages, 1987

Band 37

**Factorization of Measurable Matrix Functions**

By Georgi Semenovich Litvinchuk and
Ilya Matveyevich Spitkovskiĭ

372 pages, 1987

Band 38

**Parallel Algorithms and Architectures**

Proceedings of the International Workshop
on Parallel Algorithms and Architectures
held in Suhl (GDR), May 25–30, 1987

Edited by Andreas Albrecht, Hermann Jung,
and Kurt Mehlhorn

205 pages, 1987

Band 39

**Parametric Integer Optimization**

By Bernd Bank and Reinhard Mandel

138 pages, 1988

Band 40

**Computer Analysis of Images and Patterns 1987**

Proceedings of the II. International Conference
CAIP '87 on Automatic Image Processing,
held in Wismar (GDR), September 2–4, 1987

Edited by Leonid P. Jaroslavskiĭ,
Azriel Rosenfeld, and Wolfgang Wilhelmi

282 pages, 1987

Stability analysis and parametric optimization are of great importance both in theory and practice. Remarkable results for nonlinear problems could be obtained during the last decade. This volume is concerned with several classes of nonlinear and linear optimization problems in which some or all of the variables are additionally required to be integer numbers. The following aspects of a stability theory for the problems involved are treated:

- Existence and stability of feasible points;
- Semicontinuity properties of the extreme values, the level sets and the optimal points;
- Existence of optimal points.

Some applications of the parametric analysis to the theory of integer (linear) programming and solution concepts are presented.

---

Stabilitätsanalyse und parametrische Optimierung sind von großer Bedeutung sowohl in der Theorie als auch in der Praxis. Für nichtlineare Probleme konnten in den letzten zehn Jahren bemerkenswerte Ergebnisse erzielt werden. Dieser Band ist der Betrachtung verschiedener Klassen von nichtlinearen und linearen Optimierungsproblemen gewidmet, bei welchen einige oder alle Variable zusätzlich ganzzahlig sein müssen.
Folgende Aspekte der Stabilitätstheorie für die hier behandelten Aufgaben werden untersucht:

- Existenz und Stabilität zulässiger Punkte;
- Halbstetigkeitseigenschaften der Extremwerte, der Niveaumengen und der optimalen Punkte;
- Existenz von optimalen Punkten.

Einige Anwendungen der parametrischen Analyse auf die Theorie der (linearen) ganzzahligen Optimierung und die Konzeption von Lösungsverfahren werden diskutiert.

ISBN 3-05-500398-5
ISSN 0138-3019

MIX
Papier | Fördert
gute Waldnutzung
FSC® C083411

Zeitfracht Medien GmbH
Ferdinand-Jühlke-Straße 7
99095 Erfurt, Deutschland
produktsicherheit@kolibri360.de